UNDER THE WEATHER

Climate, Ecosystems, and Infectious Disease

Committee on Climate, Ecosystems, Infectious Disease,
and Human Health

Board on Atmospheric Sciences and Climate

Division on Earth and Life Studies

National Research Council

NATIONAL ACADEMY PRESS
Washington, D.C.

NATIONAL ACADEMY PRESS • 2101 Constitution Ave., N.W. • Washington, DC 20418

NOTICE: The project that is the subject of this report was approved by the Governing Board of the National Research Council, whose members are drawn from the councils of the National Academy of Sciences, the National Academy of Engineering, and the Institute of Medicine. The members of the committee responsible for the report were chosen for their special competences and with regard for appropriate balance.

Support for this project was provided by U.S. Environmental Protection Agency, Centers for Disease Control and Prevention, National Science Foundation, National Aeronautics and Space Administration, National Oceanic and Atmospheric Administration, U.S. Geological Survey, U.S. Global Change Research Program, and the Electric Power Research Institute. Any opinions, findings, conclusions, or recommendations expressed in this publication are those of the authors and do not necessarily reflect the views of the sponsors.

Library of Congress Cataloging-in-Publication Data

Under the weather : climate, ecosystems, and infectious disease /
National Research Council Division on Earth and Life Studies Board on
Atmospheric Sciences and Climate Committee on Climate, Ecosystems,
Infectious Disease, and Human Health.
    p. cm.
Includes bibliographical references and index.
  ISBN 0-309-07278-6
  1. Medical climatology. 2. Epidemiology. 3. Communicable diseases.
I. National Research Council (U.S.). Committee on Climate, Ecosystems,
Infectious Disease, and Human Health.
  RA793 .U53 2001
  616.9'88—dc21
                             2001001905

Additional copies of this report are available from:

    National Academy Press
    2101 Constitution Avenue, NW
    Box 285
    Washington, D.C. 20055
    800-624-6242
    202-334-3313 (in the Washington metropolitan area)
    www.nap.edu

Cover: Images on the cover were obtained from the Centers for Disease Control and Prevention "Public Health Image Library" at http://phil.cdc.gov/phil/default.asp and the NOAA Photo Library at: http://www.photolib.noaa.gov.

Printed in the United States of America

THE NATIONAL ACADEMIES

National Academy of Sciences
National Academy of Engineering
Institute of Medicine
National Research Council

The **National Academy of Sciences** is a private, nonprofit, self-perpetuating society of distinguished scholars engaged in scientific and engineering research, dedicated to the furtherance of science and technology and to their use for the general welfare. Upon the authority of the charter granted to it by the Congress in 1863, the Academy has a mandate that requires it to advise the federal government on scientific and technical matters. Dr. Bruce M. Alberts is president of the National Academy of Sciences.

The **National Academy of Engineering** was established in 1964, under the charter of the National Academy of Sciences, as a parallel organization of outstanding engineers. It is autonomous in its administration and in the selection of its members, sharing with the National Academy of Sciences the responsibility for advising the federal government. The National Academy of Engineering also sponsors engineering programs aimed at meeting national needs, encourages education and research, and recognizes the superior achievements of engineers. Dr. William A. Wulf is president of the National Academy of Engineering.

The **Institute of Medicine** was established in 1970 by the National Academy of Sciences to secure the services of eminent members of appropriate professions in the examination of policy matters pertaining to the health of the public. The Institute acts under the responsibility given to the National Academy of Sciences by its congressional charter to be an adviser to the federal government and, upon its own initiative, to identify issues of medical care, research, and education. Dr. Kenneth I. Shine is president of the Institute of Medicine.

The **National Research Council** was organized by the National Academy of Sciences in 1916 to associate the broad community of science and technology with the Academy's purposes of furthering knowledge and advising the federal government. Functioning in accordance with general policies determined by the Academy, the Council has become the principal operating agency of both the National Academy of Sciences and the National Academy of Engineering in providing services to the government, the public, and the scientific and engineering communities. The Council is administered jointly by both Academies and the Institute of Medicine. Dr. Bruce M. Alberts and Dr. William A. Wulf are chairman and vice chairman, respectively, of the National Research Council.

Acknowledgment of Reviewers

This report has been reviewed by individuals chosen for their diverse perspectives and technical expertise, in accordance with procedures approved by the National Research Council's Report Review Committee. The purpose of this independent review is to provide candid and critical comments that will assist the authors and the NRC in making the published report as sound as possible and to ensure that it meets institutional standards for objectivity, evidence, and responsiveness to the study charge. The content of the review comments and draft manuscript remain confidential to protect the integrity of the deliberative process. We wish to thank the following individuals for their participation in the review of this report:

William E. Gordon, Rice University, Houston, Texas
Nicholas Graham, Scripps Institution of Oceanography, San Diego, California
Donald A. Henderson, Johns Hopkins University, Baltimore, Maryland
Joshua Lederberg, The Rockefeller University, New York
Simon Levin, Princeton University, Princeton, New Jersey
Mercedes Pascual, University of Michigan, Ann Arbor
Roger Pielke, Jr., National Center for Atmospheric Research, Boulder, Colorado
Arthur Reingold, University of California, Berkeley
Peter B. Rhines, University of Washington, Seattle
David J. Rogers, University of Oxford, England
Mary Wilson, Harvard University, Cambridge, Massachusetts

Although the reviewers listed above provided many constructive comments and suggestions, they were not asked to endorse the conclusions or recommendations nor did they see the final draft of the report before its release. The review of this report was overseen by Lynn Goldman (Johns Hopkins University) appointed by the Division on Earth and Life Studies, and Gilbert Omenn (University of Michigan) appointed by the NRC's Report Review Committee, who were responsible for making certain that an independent examination of this report was carried out in accordance with institutional procedures and that all review comments were carefully considered. Responsibility for the final content of this report rests entirely with the authoring committee and the institution.

Preface

Over the past several years, scientists, public health officials, and policy makers have become increasingly interested in understanding how the emergence and spread of infectious diseases could be affected by environmental factors, particularly variations in climate. In September 1995 the Institute of Medicine/ National Academy of Sciences and the National Science and Technology Council held a Conference on Human Health and Global Climate Change. Following this event, an interagency discussion group met several times and decided that a more in-depth exploration of this issue was needed, and thus plans were developed for this study on climate, ecosystems, infectious diseases, and health (CEIDH).

Support for this study was provided by the U.S. Environmental Protection Agency, the Centers for Disease Control and Prevention, the National Science Foundation, the National Aeronautics and Space Administration, the National Oceanic and Atmospheric Administration, the U.S. Geological Survey, U.S. Global Change Research Program, and the Electric Power Research Institute.

The study committee, consisting of 16 people from a broad range of disciplinary backgrounds, was appointed in January 1999; see Appendix A for biographical details on the committee members. Over the course of the next 18 months, six meetings were held, where the committee received briefings from federal agency representatives, talked with experts on a wide variety of topics relevant to the study, and worked on this report. See Appendix B for a detailed list of the discussion topics and speakers at the meetings.

While this study was under way, several other assessment activities related to the issue of climate and health were being carried out, for instance, by the American Academy of Microbiology, the U.S. Global Change Research Program, and the Intergovernmental Panel on Climate Change. The CEIDH committee has followed the progress of these other activities, and in fact some committee members participated in them. However, the committee's final deliberations, and the recommendations and conclusions contained in this report, were developed independently of these other activities.

As the report title implies, this study explores the linkages among climate, ecosystems, infectious diseases, and human health. This study is global in scope; the committee considered infectious disease threats not only to the United States, but also to populations around the world. The study focuses only on the issue of infectious diseases, but it should be noted that there are many ways that climate and weather can affect human health, including the direct physical impacts of temperature extremes and severe storms, and the respiratory effects of heat-exacerbated air pollution.

An important goal of this report is to help the different groups of researchers involved in climate and infectious disease studies gain a more realistic understanding of the current capabilities and limitations of each other's fields. For instance, climatologists need to understand that epidemiological data from many parts of the world are highly limited or nonexistent, and a great deal of effort will be needed to improve this situation. In turn, epidemiologists and other health professionals need to understand the considerable uncertainties associated with many aspects of climate forecasting. Improving this mutual understanding will help ensure that future research activities are effectively designed, and that all involved have realistic expectations about the feasibility of climate-based disease early warning systems.

The primary intended audiences for this report are the scientists and program managers responsible for planning and carrying out future research on this topic. However, this issue is certainly of interest to a wider audience, and thus the committee attempted to write a report that would be accessible to people from a broad range of educational and professional backgrounds.

Contents

Executive Summary

Health and climate have been linked since antiquity. In the fifth century B.C., Hippocrates observed that epidemics were associated with natural phenomena rather than deities or demons. In modern times, our increasing capabilities to detect and predict climate variations such as the El Niño/Southern Oscillation (ENSO) cycle, coupled with mounting evidence for global climate change, have fueled a growing interest in understanding the impacts of climate on human health, particularly the emergence and transmission of infectious disease agents.

Simple logic suggests that climate can affect infectious disease patterns because disease agents (viruses, bacteria, and other parasites) and their vectors (such as insects or rodents) are clearly sensitive to temperature, moisture, and other ambient environmental conditions. The best evidence for this sensitivity is the characteristic geographic distribution and seasonal variation of many infectious diseases. Weather and climate affect different diseases in different ways. For example, mosquito-borne diseases such as dengue, malaria, and yellow fever are associated with warm weather; influenza becomes epidemic primarily during cool weather; meningitis is associated with dry environments; and cryptosporidiosis outbreaks are associated with heavy rainfall. Other diseases, particularly those transmitted by direct interpersonal contact such as HIV/AIDS, show no clear relationship to climate. By carefully studying these associations and their underlying mechanisms, we hope to gain insights into the factors that drive the emergence and seasonal/interannual variations in contemporary epidemic diseases and, possibly, to understand the potential future disease impacts of long-term climate change.

The U.S. federal agencies entrusted with guarding the nation's health and the environment, along with other concerned institutions, requested the formation of a National Research Council committee to evaluate this issue. Specifically, the committee was asked to undertake the following three tasks:

1. Conduct an in-depth, critical review of the linkages between temporal and spatial variations of climate and the transmission of infectious disease agents;
2. Examine the potential for establishing useful health-oriented climate early-warning and surveillance systems, and for developing effective societal responses to any such early warnings;
3. Identify future research activities that could further clarify and quantify possible connections between climate variability, ecosystems, and the transmission of infectious disease agents, and their consequences for human health.

There are many substantial research challenges associated with studying linkages among climate, ecosystems, and infectious diseases. For instance, climate-related impacts must be understood in the context of numerous other forces that drive infectious disease dynamics, such as rapid evolution of drug- and pesticide-resistant pathogens, swift global dissemination of microbes and vectors through expanding transportation networks, and deterioration of public health programs in some regions. Also, the ecology and transmission dynamics of different infectious diseases vary widely from one context to the next, thus making it difficult to draw general conclusions or compare results from individual studies. Finally, the highly interdisciplinary nature of this issue necessitates sustained collaboration among disciplines that normally share few underlying scientific principles and research methods, and among scientists that may have little understanding of the capabilities and limitations of each other's fields.

In light of these challenges, the scientific community is only beginning to develop the solid scientific base needed to answer many important questions, and accordingly, in this report the committee did not attempt to make specific predictions about the likelihood or magnitude of future disease threats. Instead, the focus is on elucidating the current state of our understanding and the factors that, at present, may limit the feasibility of predictive models and effective early warning systems. The following is a summary of the committee's key findings and recommendations:

KEY FINDINGS: LINKAGES BETWEEN CLIMATE AND INFECTIOUS DISEASES

Weather fluctuations and seasonal-to-interannual climate variability influence many infectious diseases. The characteristic geographic distributions and seasonal variations of many infectious diseases are *prima facie* evidence of linkages with weather and climate. Studies have shown that factors such as temperature, precipitation, and humidity affect the lifecycle of many disease pathogens and vectors (both directly, and indirectly through ecological changes) and thus can potentially affect the timing and intensity of disease outbreaks. However, disease incidence is also affected by factors such as sanitation and public health services, population density and demographics, land use changes, and travel patterns. The importance of climate relative to these other variables must be evaluated in the context of each situation.

Observational and modeling studies must be interpreted cautiously. There have been numerous studies showing an association between climatic variations and disease incidence, but such studies are not able to fully account for the complex web of causation that underlies disease dynamics and thus may not be reliable indicators of future changes. Likewise, a variety of models have been developed to simulate the effects of climatic changes on incidence of diseases such as malaria, dengue, and cholera. These models are useful heuristic tools for testing hypotheses and carrying out sensitivity analyses, but they are not necessarily intended to serve as predictive tools, and often do not include processes such as physical/biological feedbacks and human adaptation. Caution must be exercised then in using these models to create scenarios of future disease incidence, and to provide a basis for early warnings and policy decisions.

The potential disease impacts of global climate change remain highly uncertain. Changes in regional climate patterns caused by long-term global warming could affect the potential geographic range of many infectious diseases. However, if the climate of some regions becomes more suitable for transmission of disease agents, human behavioral adaptations and public health interventions could serve to mitigate many adverse impacts. Basic public health protections such as adequate housing and sanitation, as well as new vaccines and drugs, may limit the future distribution and impact of some infectious diseases, regardless of climate-associated changes. These protections, however, depend upon maintaining strong public health programs and assuring vaccine and drug access in the poorer countries of the world.

Climate change may affect the evolution and emergence of infectious diseases. Another important but highly uncertain risk of climate change are the potential impacts on the evolution and emergence of infectious disease agents. Ecosystem instabilities brought about by climate change and concurrent stresses such as land use changes, species dislocation, and increasing global travel could potentially influence the genetics of pathogenic microbes through mutation and horizontal gene transfer, and could give rise to new interactions among hosts and disease agents. Such changes may foster the emergence of new infectious disease threats.

There are potential pitfalls in extrapolating climate and disease relationships from one spatial/temporal scale to another. The relationships between climate and infectious disease are often highly dependent upon local-scale parameters, and it is not always possible to extrapolate these relationships meaningfully to broader spatial scales. Likewise, disease impacts of seasonal to inter-annual climate variability may not always provide a useful analog for the impacts of long-term climate change. Ecological responses on the timescale of an El Niño event, for example, may be significantly different from the ecological responses and social adaptations expected under long-term climate change. Also, long-term climate change may influence regional climate variability patterns, hence limiting the predictive power of current observations.

Recent technological advances will aid efforts to improve modeling of infectious disease epidemiology. Rapid advances being made in several disparate scientific disciplines may spawn radically new techniques for modeling of infectious disease epidemiology. These include advances in sequencing of microbial genes, satellite-based remote sensing of ecological conditions, the development of Geographic Information System (GIS) analytical techniques, and increases in inexpensive computational power. Such technologies will make it possible to analyze the evolution and distribution of microbes and their relationship to different ecological niches, and may dramatically improve our abilities to quantify the disease impacts of climatic and ecological changes.

KEY FINDINGS: THE POTENTIAL FOR DISEASE EARLY WARNING SYSTEMS

As our understanding of climate/disease linkages is strengthened, epidemic control strategies should aim towards complementing "surveillance and response" with "prediction and prevention." Current strategies for controlling infectious disease epidemics depend largely on surveillance for new outbreaks followed by a rapid response to control the epidemic. In some contexts, however, climate forecasts and environmental observations could potentially be used to identify areas at high risk for disease outbreaks and thus aid efforts to limit the extent of epidemics or even prevent them from occurring. Operational disease early warning systems are not yet generally feasible, due to our limited understanding of most climate/disease relationships and limited climate forecasting capabilities. But establishing this goal will help foster the needed analytical, observational, and computational developments.

The potential effectiveness of disease early warning systems will depend upon the context in which they are used. In cases where there are relatively simple, low-cost strategies available for mitigating risk of epidemics, it may be feasible to establish early warning systems based only on a general understanding of climate/disease associations. But in cases where the costs of mitigation actions are significant, a precise and accurate prediction may be necessary, requiring a more thorough mechanistic understanding of underlying climate/disease relationships. Also, the accuracy and value of climate forecasts will vary significantly depending on the disease agent and the locale. For instance, it will be possible to issue sufficiently reliable ENSO-related disease warnings only in regions where there are clear, consistent ENSO-related climate anomalies. Finally, investment in sophisticated warning systems will be an effective use of resources only if a country has the capacity to take meaningful actions in response to such warnings, and if the population is significantly vulnerable to the hazards being forecast.

Disease early warning systems cannot be based on climate forecasts alone. Climate forecasts must be complemented by an appropriate suite of indicators from ongoing meteorological, ecological, and epidemiological surveillance systems. Together, this information could be used to issue a "watch" for regions at risk and subsequent "warnings" as surveillance data confirm earlier projections. Development of disease early warning systems should also include vulnerability and risk analysis, feasible response plans, and strategies for effective public communication. Climate-based early warning systems being developed for other applications, such as agricultural planning and famine prevention, provide many useful lessons for the development of disease early warning systems.

Development of early warning systems should involve active participation of the system's end users. The input of stakeholders such as public health officials and local policymakers is needed in the development of disease early warning systems, to help ensure that forecast information is provided in a useful manner and that effective response measures are developed. The probabilistic nature of climate forecasts must be clearly explained to the communities using these forecasts, so that response plans can be developed with realistic expectations for the range of possible outcomes.

RECOMMENDATIONS FOR FUTURE RESEARCH AND SURVEILLANCE

Research on the linkages between climate and infectious diseases must be strengthened. In most cases, these linkages are poorly understood and research to understand the causal relationships is in its infancy. Methodologically rigorous studies and analyses will likely improve our nascent understanding of these linkages and provide a stronger scientific foundation for predicting future changes. This can best be accomplished with investigations that utilize a variety of analytical methods (including analysis of observational data, experimental manipulation studies, and computational modeling), and that examine the consistency of climate/disease relationships in different societal contexts and across a variety of temporal and spatial scales. Progress in defining climate and infectious disease linkages can be greatly aided by focused efforts to apply recent technological advances such as remote sensing of ecological changes, high-speed computational modeling, and molecular techniques to track the geographic distribution and transport of specific pathogens.

Further development of disease transmission models is needed to assess the risks posed by climatic and ecological changes. The most appropriate modeling tools for studying climate/disease linkages depend upon the scientific information available. In cases where there is limited understanding of the ecology and transmission biology of a particular disease, but sufficient historical data on disease incidence and related factors, statistical-empirical models may be most useful. In cases where there are insufficient surveillance data, "first principle" mechanistic models that can integrate existing knowledge about climate/disease linkages may have the most heuristic value. Models that have useful predictive value will likely need to incorporate elements of both these approaches. Integrated assessment models can be especially useful for studying the relationships among the multiple variables that contribute to disease outbreaks, for looking at long-term trends, and for identifying gaps in our understanding.

Epidemiological surveillance programs should be strengthened. The lack of high-quality epidemiological data for most diseases is a serious obstacle to improving our understanding of climate and disease linkages. These data are necessary to establish an empirical basis for assessing climate influences, for establishing a baseline against which one can detect anomalous changes, and for developing and validating models. A concerted effort, in the United States and internationally, should be made to collect long-term, spatially resolved disease surveillance data, along with the appropriate suite of meteorological and ecological observations. Centralized, electronic databases should be developed to facilitate rapid, standardized reporting and sharing of epidemiological data among researchers.

Observational, experimental, and modeling activities are all highly interdependent and must progress in a coordinated fashion. Experimental and observational studies provide data necessary for the development and testing of models; and in turn, models can provide guidance on what types of data are most needed to further our understanding. The committee encourages the establishment of research centers dedicated to fostering meaningful interaction among the scientists involved in these different research activities through long-term collaborative studies, short-term information-sharing projects, and interdisciplinary training programs. The National Center for Ecological Analysis and Synthesis provides a good model for the type of institution that would be most useful in this context.

Research on climate and infectious disease linkages inherently requires interdisciplinary collaboration. Studies that consider the disease host, the disease agent, the environment, and society as an interactive system will require more interdisciplinary collaboration among climate modelers, meteorologists, ecologists, social scientists, and a wide array of medical and public health professionals. Encouraging such efforts requires strengthening the infrastructure within universities and funding agencies for supporting interdisciplinary research and scientific training. In addition, educational programs in the medical and public health fields need to include interdisciplinary programs that explore the environmental and socioeconomic factors underlying the incidence of infectious diseases.

Numerous U.S. federal agencies have important roles to play in furthering our understanding of the linkages among climate, ecosystems, and infectious disease. There have been a few programs established in recent years to foster interdisciplinary work in applying remote-sensing and GIS technologies to epidemiological investigations. The committee applauds these efforts and encourages all of the relevant federal agencies to support interdisciplinary research programs on climate and infectious disease, along with an interagency working group to help ensure effective coordination among these different programs. The U.S. Global Change Research Program (USGCRP) may provide an appropriate forum for this type of coordinating body. This will require, however, that organizations such as the Centers for Disease Control and Prevention, and the National Institute of Allergy and Infectious Diseases become actively involved with the USGCRP.

Finally, the committee wishes to emphasize that even if we are able to develop a strong understanding of the linkages among climate, ecosystems, and infectious diseases, and in turn, are able to create effective disease early warning systems, there will always be some element of unpredictability in climate variations and infectious disease outbreaks. Therefore, a prudent strategy is to set a high priority on reducing people's overall vulnerability to infectious disease through strong public health measures such as vector control efforts, water treatment systems, and vaccination programs.

1

Introduction

Whoever would study medicine aright must learn of the following subjects. First he must consider the effect of the seasons of the year and the differences between them. Secondly he must study the warm and the cold winds, both those which are in common to every country and those peculiar to a particular locality. Lastly, the effect of water on the health must not be forgotten.

(Hippocrates, *Airs, Waters, and Places*)

A change in weather can lead to the appearance of epidemic disease. This simple observation has been appreciated since the dawn of medical science when Hippocrates taught that many specific human illnesses were linked to changes of season or temperature. Indeed, the very terms we use today for infectious diseases often preserve ancient notions of disease being caused by environmental change and other external factors. Familiar etymological examples are "influenza," which is derived from "influence"; "malaria," contracted from "mal" and "aria" (bad air); or simply "a cold," the quaintly preserved term for an upper respiratory tract infection. Perhaps the best reflection of these widespread beliefs is the colloquial phrase "under the weather," which is taken to signify a temporary illness or indisposition without other explanation.

The birth of modern microbiology and, with it, the systematic study of the epidemiology of specific microbes altered fundamental scientific concepts of disease transmission. New laboratory techniques for isolation and characterization of bacteria, viruses, and other classes of microbes identified specific agents to be the proximate cause of diseases, and control efforts accordingly shifted to a more focused scientific attack on these specific "germs." This approach appeared to be hugely successful. By the 1960s highly effective vaccines or drugs had been developed against many globally important pathogenic microbes, and many countries successfully protected their populations from disease through the use of pesticides, water treatment, and other public health measures. It appeared to be just a matter of time before the war against infectious diseases would be won, but optimism was premature. Reasons for setbacks included the rapid evolution of drug- and pesticide-resistant variants, the surprise emergence of new microbial pathogens, swift global dissemination of microbes and vectors through expanding transportation networks, and the dissipation of political will needed to sustain successful public health programs.

Given the obvious and long-appreciated linkages between climate and human health, it might seem a simple task to use climate forecasts in predictive disease models. Unfortunately, the mathematical relationships between climate and disease are neither so obvious nor so simple. Clearly establishing causal relationships between climate and disease has proven very difficult, largely for the following reasons:

• Unlike some of the other sectors commonly studied in climate impact assessments, the health sciences do not have a tradition of using predictive mathematical models or other forecasting tools.
• Infectious disease transmission patterns are affected by many factors other than climate, and the relationship between climatic variations and disease outbreaks is often mediated by ecological, biological, or societal changes.
• Data on infectious disease incidence in many parts of the world are sparse or nonexistent, which makes it difficult to develop a solid empirical understanding of climate/disease relationships.

Over the past 15 years or so, as observational data and climate modeling studies have confirmed the likelihood of a long-term global warming trend, a question that naturally arose was "what effect will this have on human health, specifically on infectious diseases?" This question has generated considerable public interest and stimulated the publication of numerous research and review papers. It has also been the focus of assessments carried out by the World Health Organization (WHO), the Intergovernmental Panel on Climate Change (IPCC), the U.S. Global Change Research Program (USGCRP), and other organizations.

Some of these publications have made claims that climate change will have wide-ranging, adverse impacts on human health. For instance, it has been projected that with warmer temperatures, the mosquito vectors that transmit malaria, dengue, and yellow fever could move northward into the United States and Europe, that development of virus and parasites would accelerate, that epidemics of diseases such as cholera will intensify with increasing sea surface temperatures, and that the emergence of new disease threats will become more common.

Studies have claimed that recent changes in infectious disease patterns (for instance, increasing malaria incidence in high altitude regions) can be linked to global warming trends. Likewise, some have concluded that interannual climate fluctuations, in particular El Niño events, have been at least partially responsible for major outbreaks of disease such as cholera and dengue; and these associations between El Niño events and disease outbreaks have been extrapolated to infer the potential impacts of long-term climate change.

At present, however, there is little solid scientific evidence to support such conclusions and few studies that take into account the full range of factors influencing pathogen transmission such as human travel and migration patterns, the collapse of public health measures in some regions, and an increase in drug resistant parasites and pesticide-resistant vectors. In addition, infectious disease experts have pointed out that mosquito vectors of malaria, dengue, and yellow fever have been in the United States for centuries, but epidemics of these diseases have vanished due to public health measures and lifestyle changes. They also make the point that humans can and do adapt to mitigate the harmful impacts of a changing climate.

A lack of consensus among the scientific community about the magnitude and relative importance of the effects of climate on infectious diseases provided one of the motivations for this NRC study. A second motivation was the level of public concern generated by press reports that have sometimes been misleading or inaccurate. Yet another motivation for the study is the hope that recent advances in climate forecasting and remote-sensing technologies can be used to provide early warnings of conditions conducive to disease outbreak. Currently, public health systems rely primarily on "surveillance and response" approaches to controlling infectious disease. It is hoped that a "prediction and prevention" approach may become more feasible with a solid understanding of the climate and ecological conditions that favor disease transmission. This challenge involves developing predictive models to make reliable disease "forecasts" and creating operational early warning systems that can effectively reduce the risk to vulnerable populations.

This report reviews the current knowledge of the relationships between climate, ecosystems, and infectious diseases, evaluates the potential for disease early warning systems, and offers recommendations about how a predictive science of infectious disease epidemiology might be realized. The report is organized as follows:

• Chapter 2 provides a historical review of how the concept of "environmental medicine" has developed over the last few centuries, and how this has shaped current perspectives on climate and infectious disease linkages.
• Chapter 3 reviews some basic concepts in climatology and infectious disease epidemiology, and gives an overview of the linkages among climate, ecosystems, and infectious diseases.

• Chapter 4 reviews the current state of understanding of how climate influences some specific diseases.

• Chapter 5 discusses the analytical approaches that can be used to study these linkages including different types of observational studies, laboratory experiments, and modeling analyses.

• Chapter 6 describes how the lessons learned from ecological studies can provide insights into the challenges of extrapolating study results from one temporal/spatial scale to another.

• Chapter 7 examines the feasibility of using climate forecasts to predict disease outbreaks and describes the different components necessary for an effective warning system.

• Chapter 8 summarizes the committee's key findings about these issues and recommendations for future research needs.

2

Climate and Infectious Diseases:
The Past as Prologue

The following review addresses the origins of environmental medicine and its legacy for our current understanding of climate and infectious disease linkages, and reviews some of the important historical milestones in the fields of meteorology and infectious disease epidemiology.

ORIGINS OF ENVIRONMENTAL MEDICINE

Many early civilizations related weather to the appearance of disease and topography to the persistence of disease in a region or population. The most lasting formulation was attributed to Hippocrates (~460 to 377 B.C.), who linked responsibility for disease to observable natural phenomena rather than to deities or demons. The seasonal appearances of particular diseases formed the basis of the Hippocratic treatise on epidemics, and many aphorisms handed down over the centuries similarly attributed various morbid conditions to weather and seasonal change (Hannaway, 1993). Hippocratic medicine focused on predicting the course and outcome of an illness through detailed observations of clinical symptoms and through associations with the way that winds, waters, and seasons appeared to make some diagnoses more likely (Smith, 1979).

Throughout the first millennium, epidemics were explained by changes in the course of the stars and other atmospheric, meteorological, and terrestrial phenomena. Popular ideas of being "under the weather" and the association of some regions with either illness or good health were used to explain the spread of disease among people who otherwise had little in common (Seargent, 1982). However, this tidy synthesis began to break down with the appearance of bubonic plague in the 1340s. The first great pandemic was preceded by a rare conjunction of three planets, which provided a convenient explanation for the extraordinary mortality. But subsequent plague epidemics were not so easily linked to meteorological or cosmological phenomena. Moreover, the specific recurrence of identical pathological features—a large swelling in particular locations on the body—linked the new disease with a few designated "contagions," diseases that were thought to be transmitted through contact (Nutton, 1983; Pelling, 1993).

On a more practical level, the extraordinary costs of plague brought a demand for novel strategies in public intervention. In order to predict and manage the local appearance of an epidemic, Italian bureaucrats in the fifteenth century designed mortality registers as a way to anticipate and monitor epidemic disease, leading to the study of temporal and geographical spread of infection. Surveillance and containment practices were invented piecemeal over the next two to three centuries, with greater attention paid to prevention and to populations at greatest risk. Plague victims and their household contacts were segregated during epidemics, which helped reinforce observations that disease was caused by local miasma (an unhealthy environment) or by contagions running rampant in the segregated population. In order to escape containment regulations, people urged physicians to distinguish plague from among the various types of buboes (i.e., swollen glands) and fevers, and this brought a new emphasis on clinical diagnosis (Cipolla, 1981; Slack, 1985; Carmichael, 1991; Pullan, 1992).

The theories of Copernicus, Galileo, Descartes, Newton, and other natural philosophers of the sixteenth through the eighteenth centuries displaced the ancient view that events on earth reflect the influence of heavenly bodies with the view that such events are subject to physical laws (Grant, 1994). The explanatory power of these physical laws led philosophers and investigators in the life sciences and medicine to search for a law-like synthesis comparable to Newton's while trying to salvage as much as possible of ancient scientific wisdom. The chosen path for the study of disease was the collection of data, both epidemiological and meteorological. From 1650 until 1850, Western science and medicine sought a comfortable synthesis of the physical and life sciences that would accommodate the vast array of new observations (Glacken, 1967; Barzun, 2000). Weather-related observations intrigued both physicians and "natural philosophers." Although histories of the development of meteorology, mathematics, and mapping often lack description of the relationship with medicine and health, the interests of many pioneers in these fields were propelled by the desire to use the scientific collection of data to understand geographic patterns of disease (Jordanova, 1979; Riley, 1987).

THE EARLY MERGER OF METEOROLOGY AND MEDICINE

The quests to understand weather, climate, and disease all posed significant methodological challenges. The Scientific Revolution brought mathematics to the study of natural phenomena, with the invention of instruments such as barometers, hygrometers, thermometers, microscopes, telescopes, water pumps, rain gauges, and devices to measure wind velocity (Frisinger, 1977). Scientific societies dating to the 1600s, the Italian Accademia del Cimento and the English Royal Society, sponsored efforts to unify and analyze meteorological and medical information. During the eighteenth century, France and Germany began to gather and synthesize local records of death to try to uncover laws of population and disease (Desaive, 1972; Hannaway, 1972). Called "statists," and their data "statistics," these early epidemiologists of the eighteenth and early nineteenth centuries expected that their accumulated medical information would inform state policy (Porter, 1986).

In the late eighteenth century, environmental medicine reached its apogee. New medical information led to broad social reforms such as the redesign of prisons, the construction of sewer and drainage systems, the early management of water resources, and the firm association of some locales and populations with increased risk of disease and death (Riley, 1987). Physicians of this era enthusiastically measured ambient temperatures, rainfall, seasonal changes in disease patterns, and many features of the natural topography, hoping to uncover an explanatory correlation between weather and disease. France even required a precise collection of meteorological data by physicians three times each day, on forms provided by the Royal Society of Medicine (Desaive, 1972; Hannaway, 1972).

In North America the interdependency of medicine and meteorology provides an especially interesting example, because the U.S. national weather bureau emerged unambiguously from the effort to link weather and disease (Cassedy, 1986). Many prominent North Americans, including Thomas Jefferson and Noah Webster, persevered in their observations of climate and disease, and the practice remained routine in U.S. military and naval services until the 1880s (Whitnah, 1965; Fleming, 1990). Continuing interest in distinctive American storms helped inspire record keeping long after European physicians concluded that weather crises could not easily be linked to epidemics (Monmonier, 1999). Because of the severity of the monsoon season in South Asia, British physicians in India also pursued climatic explanations for epidemics far longer than their European counterparts (Fein and Stephens, 1987).

Nonetheless, by the 1840s and 1850s the study of climate in Europe and North America became contested professional terrain. Early weather specialists, rather than physicians, began to collect and outline new uses for the data. Systematic collection of meteorological records led to the first American publications of theoretical meteorology (Whitnah, 1965; Kutzbach, 1979; Fleming, 1990). A decade before the Civil War, meteorologists seized the invention of the telegraph and envisioned storm warnings to reduce the losses from floods and cold waves. Aided by a network of telegraph operators and a central collection station at the Smithsonian Institution, scientists pursued the goal of using meteorology for prediction and forecasting, although physicians still collected the weather observations that provided the data for meteorological publications. The military engulfed meteorology, and the first U.S. National Weather Service was created in 1869 as part of the Army's Signal Service. By 1880 meteorology's early origins in medicine were all but invisible, and medical study of epidemics turned away from prediction toward pursuit of disease agents in laboratory-based investigations.

METEOROLOGY BECOMES AN INDEPENDENT DISCIPLINE

Frederik Nebeker categorized the emerging communities of meteorology from the mid-nineteenth century until the end of World War I into observers, forecasters, and theoreticians (Nebeker, 1995). Of the three traditions, the observers, an empirically-based, observation-driven community of climatologists, maintained the most explicit links to the older tradition of environmental medicine.

Historians of meteorology date the breakthrough in modern meteorology to the period during and immediately after World War I, when the demands of war increased interest in weather forecasting. The great pioneer of physics-based meteorology, Vilhelm Bjerknes, founded a school in Norway that was one of the first in theoretical meteorology to accept practical weather forecasting as a necessary application of science to meet societal goals (Friedman, 1989). Bjerknes and other twentieth-century meteorologists created the explicitly war-related terms of "fronts" and "air masses" to describe larger regional atmospheric processes.

The prospect of war directed ever greater resources into meteorological research during the twentieth century. The enormous century-long collection of observational meteorological data fueled much of the effort to develop models for making weather predictions. Lewis Frye Richardson, a young mathematician and physicist, devised mathematical equations that would transform atmospheric data from one place and time into a prediction of the weather six hours later. His predictions erred by two orders of magnitude, but his 1922 book, *Weather Prediction by Numerical Process,* nonetheless represented a breakthrough on the path to theory-based forecasting. Meteorologists responded to Richardson's early attempts to predict weather with demands for more weather stations, more accurate instruments, better training for observers, standardization, and international collaboration (Ashford, 1985).

A meteorology program organized by John von Neumann in 1946 revised the dynamic equations of Richardson and applied the new computer technology to weather forecasting models. In 1950 Neumann's group used the ENIAC computer to make the first numerical prediction of weather. The invention of the digital computer lifted the constraint on calculations (Aspray, 1990). With the aid of the computer, however, meteorologists were faced with the problem that very small differences in the initial dataset could yield extremely large differences in calculated predictions. In response they made refinements to the models and equations as understanding of atmospheric physics advanced. These advances led to unprecedented predictive success and a new mathematics, described by one of its pioneers, meteorologist Edward Lorenz. Lorenz described the theoretical problem of data in a presentation he made before the American Association for the Advancement of Science in 1972 with the memorable title "Predictability: Does the Flap of a Butterfly's Wings in Brazil Set Off a Tornado in Texas?" His application of the mathematics of chaos to weather forecasting demonstrated that prediction in chaotic systems was not impossible, just difficult. Lorenz also offered a corollary: "If the flap of a butterfly's wings can be instrumental in generating a tornado, it can equally well be instrumental in preventing a tornado." In this model, prediction could also lead to intervention (Lorenz, 1993).

MEDICAL ENVIRONMENTALISM WITHOUT METEOROLOGY

Epidemiology emerged as a new science-oriented discipline during the 1840s and 1850s. Much of the success of early epidemiologists during the mid-nineteenth century came from studies on specific problems of purifying airs, waters, and places. Others sought to link poverty, hunger, criminality, or occupation to health problems at the population level. Many "miasmatists" or "environmentalists" also blamed disease on the presence of microscopic particles or polluting chemicals (Kunitz, 1987). When analyzing causes for human diseases, scientific medicine of the period from 1850 to 1920 tended to concentrate either on isolation of substances that caused particular diseases (microorganisms, poisons, and toxins) or on the association of disease with place, time, group, season, geographical climate, and more subjective concepts of race and social class. Some epidemiologists even pored over historical compilations of epidemics, famines, floods, and other natural disasters looking for patterns and cycles in much the same fashion as did contemporary meteorologists (Walford, 1879; Hirsch, 1886; Creighton, 1969; Corradi, 1972).

The first "revolution" in epidemiology came after 1882 with Robert Koch's elaboration of the germ theory of disease (Winslow, 1980). Younger, statistically knowledgeable epidemiologists accepted the fundamental determinism of germ theory—that a particular microorganism would be found responsible for any given disease. Most Europeans adopted the gospel of urban water purification, followed by the laboratory evidence and logical proof of the germ theory (Evans, 1978). John Snow, a hero of mid-twentieth-century epidemiologists, demonstrated an association between cholera infection and the water delivered by one pump in London in 1854 (Vandenbroucke, 1989). However, Robert Koch did not actually prove the waterborne nature of cholera until 30 years later. Tracing an epidemic back to Calcutta, India, Koch first identified the *vibrio* or "comma" bacillus (from its shape) in the intestines of cholera victims. Because Koch could not reproduce infection in an animal model according to his "postulates," he devised a statistical geographical proof of the transmission of cholera (Coleman, 1987).

Table 2-1 presents an outline of twentieth-century epidemiology constructed by Susser and Susser (1996). In this view the post-World War II era witnessed the "second revolution" in epidemiology, a reordered focus on environmental risk factors leading to chronic diseases. "Environmental" by the 1950s included behaviors or habits, along with the old "airs, waters, and places." Notable studies included the association of tobacco smoking with lung cancer and of particular diets and lifestyles with cardiovascular mortality (Susser, 1985). Susser attributed this revolution primarily to Austin Bradford Hill's assimilation of modern statistical methodology to the study of chronic diseases. Less acknowledgment was made of "popular epidemiology" that focused on the health effects of toxins and industrial pollutants in the environment. Yet this aspect of public health received widespread public attention, especially following the publication of *Silent Spring* by naturalist Rachel Carson (1962).

The firm commitment to statistical methodology and study design has been interpreted as a sign of maturity in the field of epidemiology. With the revisiting of older ideas on the correlations between climate and health, epidemiologists have recently begun to reexamine their disciplinary objectives (Susser, 1986; White, 1991; McMichael, 1994; Susser, 1998). Contemporary epidemiology appreciates the "web of causation" that underlies most public health issues and maintains a strong focus on modeling the complex relationships among multiple risk factors. Yet as noted by Krieger (1994), the field still lacks an "ecosocial" framework that embraces population-level thinking and that truly integrates social and biological understandings of health, disease, and well-being.

TWENTIETH-CENTURY TELECONNECTIONS

A critical advance in twentieth-century study of climate and weather came from Jacob Bjerknes, who linked the Southern Oscillation (shifts in atmospheric pressure in the eastern South Pacific and Indian oceans) to the periodic El Niño phenomenon (Quinn and Neal, 1992; Glantz, 1996). Recent success in forecasting the 1997/1998 El Niño has fueled the optimistic outlook for predictive meteorology and renewed interest in the history of weather events. Over the last half-century, significant collaborative enterprises between historians and meteorologists have produced a sustained inquiry into regional and global climate variability during the prehistoric and historical past (Wigley et al., 1992). Also, the last 40 or 50 years have witnessed unprecedented interdisciplinary collaboration between such fields as meteorology and oceanography, and widespread public education about weather and climate phenomena. Over this same time period, the epidemiological community has solidified its professional and disciplinary independence from the medical and public health communities and has built the statistical and

TABLE 2-1 Three Eras in the Evolution of Modern Epidemiology

Era Approach	Paradigm	Analytical Approach	Preventive
Sanitary statistics (first half of nineteenth century)	Miasma: poisoning by foul emanations from soil, air, and water	Demonstrate clustering of morbidity and mortality	Drainage, sewage, sanitation
Infectious disease epidemiology (late nineteenth century through first half of twentieth century	Germ theory: single agents relate one to one to specific diseases	Laboratory isolation and culture from disease sites; experimental transmission and reproduction of lesions	Interrupt transmission (vaccines, isolation of the affected through quarantine and fever hospitals, and ultimately antibiotics)
Chronic disease epidemiology (later half of twentieth century)	Black box: exposure related to outcome without necessity for intervening factors or pathogenesis	Risk ratio of exposure to outcome at individual level in populations	Control risk factors by modifying lifestyle (diet, exercise, etc.) or environment (pollution, passive smoking, etc.)

SOURCE: Susser and Susser, 1996

mathematical sophistication required to begin to address complex issues, such as the potential health impacts of climate change.

In broadly simplified terms, the historical objective of modern medicine and public health has been to eliminate illness, starting with the identification of causative agents. Clinical medicine and epidemiology emphasize the identification of disease causes and treatment rather than prediction of future disease outbreaks. In contrast, modern meteorology has focused on prediction, to offset the most deleterious consequences of weather events. Unlike medicine, the primary goal of meteorology is not to "cure" severe weather, that is, to retrain the path of a tornado or hurricane. However, recent debate about possible anthropogenic contributions to long-term climate change has raised the question of whether human interventions to delay or reduce the magnitude of this change should be sought.

3

Linkages Between Climate, Ecosystems, and Infectious Disease

This chapter explores the basic processes underlying the linkages between climate, ecosystems, and infectious disease. In the first two sections, background concepts related to climate and to infectious disease dynamics are reviewed to provide a foundation for those who are not already familiar with these topics. The third section provides an overview of the different ways that climate can influence the emergence and transmission of infectious disease agents, and the fourth section reviews other factors that must be considered as part of a comprehensive understanding of disease dynamics.

WEATHER AND CLIMATE: BACKGROUND CONCEPTS

Climate Variability

The distinction between weather and climate is not always appreciated but is important to understand in the context of the topics discussed in this report. Weather refers to the day-to-day state of the atmosphere, characterized by meteorological factors such as temperature, humidity, precipitation, and winds. In contrast, climate refers to average meteorological conditions over a specified time period (usually at least a month), which may include information about the frequency and intensity of extreme events and other statistical characteristics of the weather. Climate varies naturally over a wide range of spatial and temporal scales.

Spatial climate variability includes well-known latitudinal and altitudinal temperature gradients. For example, under typical conditions in mountainous terrain, the average surface air or soil temperature decreases by about 6.5°C for every 1,000-meter increase in elevation, and along an equator-to-pole gradient a distance of 1,000 kilometers corresponds to an average surface temperature change of about 5°C. Superimposed on these large-scale gradients are more complex regional patterns of temperature, precipitation, storm frequency, and so forth, and on very fine geographic scales, microtopographic changes can generate large differences in surface temperature and soil moisture.

Temporal climate variations are most obviously recognized in "normal" diurnal and seasonal variations. The amplitude of the diurnal temperature cycle at most locations is typically in the range of 5 to 15°C. The amplitude of seasonal climate variability is generally larger than that of the diurnal cycle at high latitudes and smaller at low latitudes. Normal annual cycles are modulated by interannual variations in average seasonal conditions. Interannual variations in the mean annual surface air temperature at specific locations are typically on the order of a degree Celsius. Interannual variations in precipitation are more substantial, and can be comparable to the average annual value. Years of research on these types of seasonal-to-interannual variations have uncovered a set of commonly recurring pressure and wind patterns that are termed "modes" of climate variability. The best known of these modes is the El Niño/Southern Oscillation (ENSO) cycle. The irregular cycling back and forth between warm (El Niño) and cold (La Niña) phases of the ENSO cycle in the equatorial Pacific results from a complex interplay between the strength of surface winds that blow westward along the equator and subsurface currents and temperatures. El Niño events (like the very strong one that prevailed in 1997-1998) are marked by higher than average pressure over the western Pacific and lower than average pressure over the eastern Pacific.

ENSO, however, is not the only common mode of climate variability. The Pacific North American Oscillation and North Atlantic Oscillation (thought by some researchers to be part of a larger Arctic Oscillation) are also well researched and are routinely discussed in the climate research literature, although these modes are not yet as familiar to non-specialists. Much of current climate research has turned to the examination of climate variability modes with periods of 10 or more years. North America appears to have a particularly large decadal signal.

On shorter timescales the picture is less clear, but researchers have found "intraseasonal" modes of climate variability with periods of roughly one to two months. These intraseasonal modes are primarily identified by movements of large-scale rainfall patterns in the tropical Indian and Pacific oceans; they appear to interact with ENSO, but their relationships are not well understood. There are some studies suggesting that the tropical intraseasonal variability may have an influence on mid-latitude monthly and seasonal climate.

Finally, climate also varies on very long timescales, which cannot be observed directly but are inferred through a variety of "proxy" records in ice cores,

Box 3-1
Global Impacts of ENSO

The El Niño/Southern Oscillation (ENSO) cycle is one of the Earth's dominant modes of climate variability. The ENSO cycle affects climate around the globe, but only a few regions have consistent documented impacts (Figure 3-1). During warm episodes in the tropical Pacific Ocean (El Niño events), abnormally heavy rainfall over the warm Pacific waters extends into coastal Ecuador and northern Peru, while drought conditions often prevail in parts of Australia, Malaysia, Indonesia, Micronesia, Africa, northeast Brazil, Central America, and tropical Africa. The impacts of El Niño over the United States vary from one event to the next, and they tend to be restricted to winter and early spring. On average, Alaska, the Pacific Northwest, and the Northern Plains tend to be mild and dry, while the South tends to be wet. Cold years in the tropical Pacific (La Niña events) tend to be marked by the opposite kinds of climate phenomena than in the above region.

Predictions of these ENSO-associated regional anomalies are generally given in probabilistic terms, because the likelihood of seeing any projected anomaly varies a great deal from region to region and also varies with the strength and specific configuration of the equatorial Pacific sea surface temperature anomalies during the season of interest. The term "teleconnections" is often used to describe the statistical relationship between the ENSO cycle and rainfall or temperature anomalies observed in a particular geographical location.

tree rings, and numerous other means. On century-to-millennial scales, climate changes such as the "little ice age" occur; and over the past approximately million years the global climate record is characterized by larger glacial-interglacial transitions, with multiple periodicities of roughly 20,000, 40,000, and 100,000 years, associated statistically with the effects of Earth-Sun orbital variations. The amplitudes of these transitions are on the order of 5 to 10°C and are accompanied by large extensions and retreats of polar and glacial ice.

Climate Change

Since the beginning of the Industrial Revolution, emissions of greenhouse gases such as CO_2, CH_4, and N_2O have been rising due to the growing global population, increasing per capita energy consumption (primarily through the burning of fossil fuels), and land uses such as deforestation and agriculture. These greenhouse gases increase the natural atmospheric trapping of infrared (heat) radiation, a process known as radiative forcing, which results in an enhanced greenhouse effect.

Observational evidence indicates that Earth's global mean temperature has been rising over the last century, particularly during the past 20 years (NRC, 2000). Most climate models project that global temperatures will continue to increase throughout the twenty-first century, with warming of several degrees

23

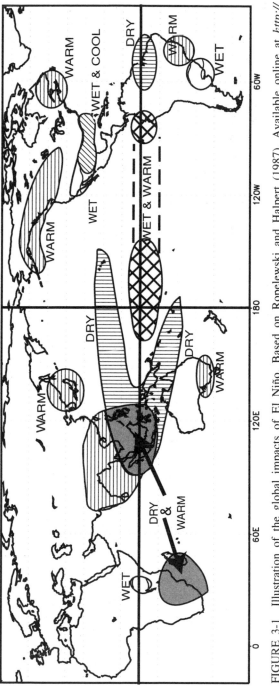

FIGURE 3-1 Illustration of the global impacts of El Niño. Based on Ropelewski and Halpert (1987). Available online at *http://www.cpc.ncep.noaa.gov/products/analysis_monitoring/impacts/warm.gif.*

Celsius by 2100. Perhaps more important than mean global temperature changes, however, is understanding how climate change will affect the "spatial geography" of climate and expressions of other modes of climate variability. The actual effects of global warming on regional-scale temperature, precipitation, and humidity are expected to vary widely. There has been speculation that the long-term global warming trend could affect the occurrence or intensity of El Niño events, but the observational climate record is too short to determine if such changes are indeed occurring. The reader is referred to IPCC (2001) for a more extensive discussion of issues related to global climate change.

In addition to the impacts of greenhouse gases, there are other anthropogenically driven processes that can affect climate on local and regional scales. Industrial emissions can greatly enhance the natural background concentration of aerosols in the atmosphere, and these particles can exert either a net warming or cooling effect, depending on numerous factors such as the particles' chemical and physical properties, the height of the aerosol layer, and the albedo of the surface beneath. Aerosols can also augment the number and properties of droplets within clouds, which in turn affect temperature and precipitation. Recently, satellite data have revealed that plumes of aerosol pollution can have a widespread influence on continental precipitation by reducing cloud particle size (Toon, 2000; Rosenfeld 2000).

Changes in land use and vegetation cover can also affect climate over a wide range of spatial scales. Vegetation can, of course, provide shade for the ground underneath, but it can also affect regional precipitation patterns, since the water vapor "exhaled" by forests is a significant source of clouds and rainfall. A modeling study by Pielke et al. (1999) estimated that loss of the South Florida Everglades over the last century has decreased rainfall in the region by about 10 percent. Likewise, there is concern that deforestation-induced drought may be occurring in the Amazon and other parts of the tropics. Another land-use impact is the "urban heat island" effect, in which cities can be up to 12°C warmer than surrounding areas due to the extra heat absorbed by asphalt and concrete and the relative lack of vegetation.

Weather and Climate Forecasting

The skill of routine weather forecasting has increased dramatically since the middle of the twentieth century, commensurate with advances in computer systems and meteorological observations. Forecasts are based on the fundamental laws of physics, including Newton's laws for fluids combined with conservation of mass and momentum. Numerical weather models start with observations of the current atmosphere and integrate forward in time. These forecasts are often defined as solutions to "initial value" problems. Because the atmosphere is an inherently chaotic system, slightly different initial conditions can produce profoundly different forecasts. Theoretical studies indicate that because of the growth of errors in describing the initial state of the atmosphere, accurate weather forecasts are not possible beyond a period of about two weeks (Lorenz, 1982).

Entirely different modeling approaches are required for seasonal climate forecasting. Early attempts to predict climate variations on monthly to seasonal timescales relied almost exclusively on statistical analysis of the past record. These empirical/statistical climate prediction techniques produced marginally skillful seasonal forecasts, but until recently the source of the forecast skill was not well understood, and there was little basis to evaluate whether one seasonal forecast would be more or less accurate than another.

Not until the 1980s did climate scientists begin to capitalize on the premise that the physical basis for seasonal climate modeling does not rest solely in the atmosphere. Seasonal climate variability is now understood to be a manifestation of complex interactions between the atmosphere and underlying surface, primarily the world oceans. The coupling between the oceans and the atmosphere at seasonal timescales was first discovered through study of the El Niño Southern Oscillation.

The current approach to numerical computer modeling of seasonal climate (e.g., Mason et al., 1999) is through the solution of what are known as boundary value problems, where the mean state of the atmosphere is coupled to the mean state of the lower boundary, especially the ocean. Atmospheric models are run with a number of initial conditions for the same boundary conditions. The resulting set of forecasts is called an "ensemble," and seasonal climate predictions are often given by the mean of ensemble forecasts. The confidence in these forecasts is sometimes gauged by the spread among the ensemble members. Given the nature of the seasonal prediction problem and the tools at hand, all such predictions are inherently probabilistic. Seasonal prediction also often incorporates empirical and statistical tools in addition to numerical model output (Barston et al., 1999; Goddard et al., in press). Typically predictions are expressed in terms of terciles (i.e., the probability that predicted seasonal rainfall or temperature will be in the upper, middle, or lower third compared to the historical record [Figure 3-2]).

Most modeling and statistical efforts have focused on the equatorial Pacific Ocean basin, but there is good evidence that the Atlantic and Indian oceans also exert an influence on the atmosphere, albeit more locally constrained to the areas surrounding those basins. The ocean dynamics and coupling between the oceans and atmosphere in the Atlantic and Indian oceans are not as well understood, but there are research activities under way to improve this understanding. There is an expectation that improved seasonal forecasts will result once the Atlantic and Indian oceans' variability is better understood and modeled.

While ENSO research has shown the profound influence of the ocean surface on seasonal climate, all atmosphere/surface boundaries may be important, including those over land. Land/atmosphere interactions may be particularly significant in influencing the local climate. There have been major research

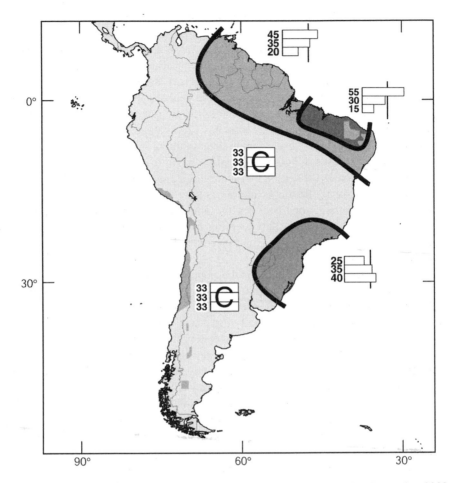

FIGURE 3-2   Example of a seasonal forecast.   A forecast for October-December 2000 (produced in September 2000) showing the percentage likelihood that the seasonal precipitation in particular regions will be above normal (top box), normal (middle box), or below normal (bottom box), compared to the mean for the period 1961 to 1990.   "C" indicates that there is no basis for favoring any particular category.   Source: International Research Institute for Climate Prediction, available online at *http://iri.ldeo.columbia.edu/ climate/forecast/net_asmt.*

efforts in the past several years to monitor, model, and predict land/atmosphere interactions. The successful incorporation of coupled land surface models into seasonal climate forecast models is expected to enhance prediction skill.

As described previously, coupling of the worlds' oceans with the global atmosphere provides a basis for climate prediction on seasonal timescales. For shorter time periods though, the influence of the ocean surface is overwhelmed by the "noise" introduced by weather. Thus, there is the conundrum that the tools for day-to-day weather prediction are constrained to less than two weeks, while the tools for climate prediction generally only work well on seasonal or longer timescales. Coupled ocean/atmosphere/land models may improve the ability of climate models to make predictions on shorter than seasonal times-cales, and more sophisticated statistical techniques may improve the use of at-mospheric weather models beyond 10 to 14 days. Decreasing this relative mini-mum in predictive capability will be one of the great scientific challenges for climate prediction research in the coming years.

Climate Change Projections

General circulation models (GCMs) represent the large-scale circulations and interactions of the atmosphere in three-dimensional form, and they are the primary tools used to make projections about how the Earth's climate will change in response to the anthropogenic addition of greenhouse gases and aerosols. Recently, scientists have begun to couple GCMs with similar models of the oceans and the biosphere in an effort to better understand the interactions within the Earth-atmosphere system and the subsequent impacts on climate.

Although GCMs are growing increasingly sophisticated, there are still sev-eral important sources of uncertainty to consider when using these models to assess possible future climatic changes. While all GCMs project increasing temperatures on a global scale, local and regional-scale projections are often very different between models. Currently GCMs have a resolution of about 200 to 500 kilometers, yet many critical physical processes, particularly those related to clouds and precipitation, take place on much smaller scales and thus must be crudely parameterized in the models. Various techniques have been developed for "downscaling" climate model simulations, such as regional climate modeling and statistical downscaling (Giorgi and Mearns, 1991; Kattenberg et al., 1996). While these techniques have steadily improved on the simulation of regional climate in the past 10 years, regional-scale projections of climate remain highly uncertain.

The uncertainty in predicting future climate results equally from uncertainty in the modeling of climate itself and uncertainty in predicting emissions of green-house gases and aerosols. The latter source of uncertainty depends on projec-tions of population growth, economic growth, and technological changes, among others. Since it is impossible to make quantitative predictions about all of these factors, a common approach is to develop a set of scenarios representing a plau-sible range of future developments.

Box 3-2
Terminology: Forecast/Projection/Prediction

There can be subtle but important distinctions in the standard use of common scientific terms among different disciplines, for example, "prediction," "forecast," and "projection." The terms forecast and prediction, which both refer to a statement about future events, are often considered to be synonymous. Meteorologists, however, tend to make the distinction that a prediction is the result of a single numerical model, while a forecast comes from a synthesis of a number of predictions. Future estimates of long-term climate change are usually discussed in terms of "projections," which are generally considered to be less certain than predictions or forecasts. Projections are based on scenarios of possible future changes (in population, economic growth, technological development, etc.) with no specific probability associated with any of the scenarios.

INFECTIOUS DISEASE DYNAMICS: BACKGROUND CONCEPTS

Epidemiological Terminology

An infectious, or communicable, disease is an illness caused by a specific infectious agent that is transmitted from an infected person, animal, or reservoir to a susceptible host. Numerous terms are used to describe the occurrence of infectious diseases in a defined population at a particular point in time. For instance, disease "prevalence" refers to the proportion of people with disease in a particular population, while disease "incidence" describes the number of new cases appearing in a population during a particular interval of time.

A disease "epidemic," often popularly described as an "outbreak," refers to an excess of cases beyond that which normally occurs in a particular region and at a particular time of year. One or two locally acquired cases of a disease that had not recently been observed in a region may be considered an outbreak. In contrast, a large number of cases in a particular area may not be considered an epidemic unless there were more than had occurred historically during the same time of year. Thus, the number of cases that constitute an epidemic or outbreak will vary with each disease, location, and season. A further complication is the fact that numerous methods are used to estimate disease incidence, not all of which are highly accurate. For instance, relying on the number of hospital visits or clinically diagnosed cases could significantly undercount the incidence of diseases that are largely asymptomatic or that have symptoms shared with other diseases.

As distinct from epidemic, the term "endemic" refers to the constant presence of a disease in a region at a roughly steady incidence during a defined period. In some cases, a disease may be "hyperendemic," that is, constantly present at a high level of incidence. Endemic diseases can also occur as periodic (often seasonal) outbreaks with varying numbers of cases depending on the intensity of transmission in a particular area, the geographic range of transmission, and the length of the transmission season.

"Emerging" diseases may be caused by a variety of pathogens that are considered to be undergoing change in particular ways. Emerging diseases can be characterized by at least one of the following criteria: (1) diseases caused by a new, previously unknown agent or syndrome; (2) disease symptoms that are more severe and/or more difficult to treat; (3) increased disease incidence in a region; and (4) widening global distribution.

Disease emergence can be due to a variety of causes, including increasing global traffic of goods and people, changes in human behavior and demographics, or a breakdown in public health measures. Other frequently identified factors in disease emergence (especially in outbreaks of previously unrecognized diseases) are climatic or ecological changes that place people in contact with a natural reservoir or host of an infection, by either increasing proximity or creating conditions that favor an increased population of the microbe or its natural host (NRC, 1992; Wilson et al., 1994; Morse, 1995).

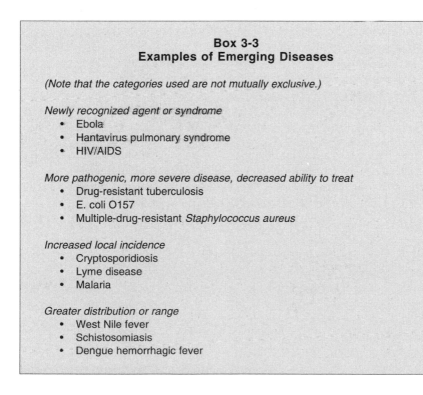

Box 3-3
Examples of Emerging Diseases

(Note that the categories used are not mutually exclusive.)

Newly recognized agent or syndrome
- Ebola
- Hantavirus pulmonary syndrome
- HIV/AIDS

More pathogenic, more severe disease, decreased ability to treat
- Drug-resistant tuberculosis
- E. coli O157
- Multiple-drug-resistant *Staphylococcus aureus*

Increased local incidence
- Cryptosporidiosis
- Lyme disease
- Malaria

Greater distribution or range
- West Nile fever
- Schistosomiasis
- Dengue hemorrhagic fever

In some cases the emergence of new infectious disease organisms reflects the genetic adaptation of viruses and bacteria to changes in the external environment (e.g., ecological conditions, climate) or the internal environment of the host (e.g., immunity acquired in response to a previous exposure to the disease organism). The illness caused by these new disease strains may have increased severity, such as hemolytic uremic syndrome caused by new strains of toxigenic *Escherichia coli*, and toxic shock syndrome caused by new strains of group A streptococci. This enhanced virulence is probably the result of genetic events and evolutionary selection (Krause, 2001).

Modes of Disease Transmission

Classifying the different modes of transmission of infectious microbes permits examination of common attributes that may be used to evaluate how climate influences the distribution of disease incidence. First, most infectious diseases can be classified as either anthroponotic or zoonotic. Anthroponotic diseases are caused by microorganisms that normally exist in a transmission cycle involving only humans. Zoonotic diseases occur when microbes that are normally transmitted among nonhuman hosts are transferred to humans. A few microbes are regularly transmitted between and among humans and animals and may cause disease in both. Each of these categories can be further classified according to whether transmission occurs directly or indirectly.

Directly transmitted anthroponoses involve two components (microbial agent, human host) and are spread among people through contact or close association. For instance, the spread of microorganisms such as rhinovirus or *Streptococci* bacteria generally occurs via bodily fluids. Transmission of microbes such as tuberculosis or measles, for example, occurs via aerosolized droplets, fecal material, fomites, or other particulates. *Indirectly transmitted anthroponoses* involve three components (agent, vector, and human). Examples include many widespread diseases caused by microbes that are transmitted via arthropod vectors, such as dengue and malaria, which are transmitted by mosquitos.

Directly transmitted zoonoses involve three components (agent, reservoir, and human) and are spread by aerosolized particles or body fluids. People do not often experience this type of infectious contact with animals; rabies is an example of the relatively few diseases of this category. *Indirectly transmitted zoonoses,* which involve four components (agent, vector, reservoir, and human) infect humans via water, soil, feces, or arthropod vectors. There are numerous examples of these types of transmission cycles, including aerosol-borne hantaviruses, water-borne cryptosporidiosis, sandfly-transmitted leishmaniasis, flea-vectored plague, tick-associated Lyme disease, and mosquito-borne Rift Valley fever. Finally, there are several infectious diseases that are not communicable but that are acquired from the environment, such as Legionnaires, tetanus, and coccidioidomycosis.

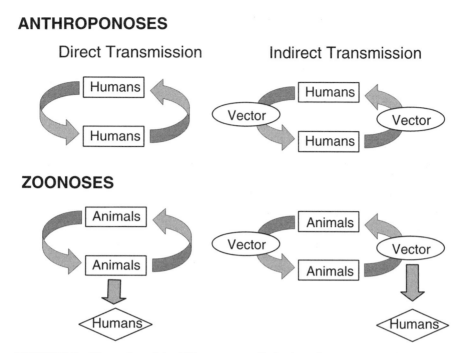

FIGURE 3-3   Illustration of the different types of infectious disease transmission cycles.

These basic categories of transmission are illustrated in Figure 3-3. The impacts of various climate factors on transmission will largely depend on the number of variables that characterize a pathogen's life cycle and the climate sensitivity of each of these variables. Directly transmitted anthroponoses, with the fewest variables and simple links, generally appear to be the least sensitive to climate influences. On the other hand, vector-borne zoonoses, with many environmentally sensitive links, may be highly influenced by climate. At the same time, this larger number of climate-sensitive links makes it more difficult to forecast how climatic changes will alter risk.

The SEIR Framework

At the heart of many studies of infectious disease transmission dynamics is an SEIR modeling framework, which depicts the different states in the progression of a disease through a population: the proportions of individuals susceptible to infection (S); the proportion of people exposed to an infectious agent but not yet infectious (E); the proportion that are actually infectious (I); and those who are removed from the population of interest (R) either because they recover from the infection and are immune or because they die.   A typical SEIR model is illustrated in Figure 3-4.

Usually this type of model consists of a series of differential equations that describe the rate of change from one state to another as a function of the relative proportions in each state and of numerous other disease- and population-specific parameters that are determined empirically from laboratory and field observations. For example, the change in the proportion of individuals in the susceptible state is determined by the initial number of susceptibles, by birth and death rates, and by the rate of contact between susceptible and infected individuals multiplied by a factor that reflects how likely this contact is to result in disease transmission.

The SEIR framework reflects the fact that disease dynamics are affected by many factors unique to a particular population, including the population size and density, demographics, connectivity patterns, and immunity levels. Infection of an individual may result in death, chronic infection, or recovery with immunity. At a population level, in some disease systems the gradual loss of widespread immunity due to the birth of susceptibles results in cyclical patterns of disease as

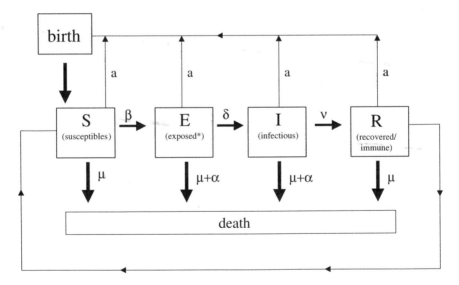

*infected but not infectious

FIGURE 3-4   Diagram of the SEIR framework used in modeling the transmission of disease agents. Definition of parameters: a = host per capital reproductive rate; μ = host per capita death rate; β = transmission coefficient; α = host per capita death rate from being infected; $1/\gamma$ = average duration of immunity; $1/\delta$ = average duration of latent period; $1/v$ = average duration of infection.

the population oscillates between resistance and receptiveness to new introductions of a pathogen. For a detailed discussion of these concepts and examples of how the SEIR framework is used in infectious disease modeling, see Mollison (1995) and Anderson and May (1992).

As discussed in Chapter 5, the SEIR modeling approach can lend itself to exploring the potential effects of climate and ecological changes on disease dynamics. In the simplest models of infectious diseases, the rates of change from one state to another in the SEIR model are assumed to be constant, but under changing environmental conditions, each rate is itself a variable. For example, the transmission rate from infected to uninfected hosts can be considered as a variable that is influenced by such factors as changing patterns of demographics and spatial aggregation in hosts and changing reproductive rates and genetics of pathogens. The effects of environmental change can be explicitly incorporated if the SEIR model is embedded in a larger system of dynamical equations that describe the relationships between the parameters of the SEIR model and climatic influences. These relationships can be derived either empirically or from first-principle mechanistic approaches.

WEATHER/CLIMATE INFLUENCES ON INFECTIOUS DISEASES: AN OVERVIEW

Disease agents and their vectors each have particular environments that are optimal for growth, survival, transport, and dissemination. Factors such as precipitation, temperature, humidity, and ultraviolet radiation intensity are part of that environment. Each of these climatic factors can have markedly different impacts on the epidemiology of various infectious diseases. Examples of typical environmental conditions and their effects on various infectious diseases are given in Table 3-1. Climate can directly impact disease transmission through its

TABLE 3-1   Examples of Diseases Influenced by Environmental Conditions

Environmental Condition	Disease Favored	Evidence
Warm	Malaria, Dengue	Primarily tropical distribution, seasonal transmission pattern
Cold	Influenza	Seasonal transmission pattern
Dry	Meningococcal meningitis, Coccidioidomycosis	Associated with arid conditions, dust storms
Wet	Cryptosporidiosis, Rift Valley fever	Associated with flooding

effects on the replication and movement (and perhaps evolution) of disease microbes and vectors. Climate can also operate indirectly through its impacts on ecology or human behavior. These different mechanisms are discussed in further detail below.

Direct effects on microbial replication rate. Infectious microorganisms have a replication rate proportional to the ambient temperature. Human body temperature is sufficient for replication of microbes causing human diseases, but ambient temperatures may drop to well below the critical threshold for replication. Malaria, dengue, and other vector-borne diseases are all caused by temperature sensitive microorganisms that replicate in mosquitos, flies, ticks, or other cold-blooded arthropods. The period of time required for microbe replication in the vector species may vary dramatically with the ambient temperature, and below certain temperature thresholds, replication may cease altogether (Gillet, 1974; Molineaux, 1988). Although less well documented, it appears that some infections of the human mucosal surfaces and skin are also sensitive to regional body temperatures. For instance, some respiratory viruses grow preferentially in the upper airways where the cells are a bit cooler than core body temperatures, and bacteria such as leprosy grow preferentially in the cooler tissues of the extremities (Robard, 1981). Waterborne diseases such as cholera also are temperature-sensitive, with a minimum temperature required for replication in the environment.

Direct effects on microbial movement. In order to be transported over the relatively large distances from one host to another, many microbes must be passively borne through moving air or water. Some pathogenic microbes, such as that causing coccidioidomycosis, are picked up from the soil and carried in dry, dusty winds (Smith et al., 1946; Schneider, 1997). Others, such as crypto-sporidiosis, may be washed by heavy rains into reservoirs of drinking water (Alterholt et al., 1998).

Direct effects on movement and replication of vectors and animal hosts. The geographic distribution of many arthropod vectors is limited by minimum and maximum temperatures, humidity, and the availability of breeding sites (which, for mosquitos, is often tied to precipitation levels). Meteorological variables may also affect the timing of the vector's life cycle and thus the rate at which it can transmit disease agents. For instance, as temperatures increase over a certain threshold, blood-feeding arthropods often increase their biting frequency and reproduction rates (Gillet, 1974; Shope, 1991; Bradley, 1993). Similarly, in the case of epizootic diseases, the abundance of the primary host (generally another vertebrate species) may depend on climate variables and affect the incidence of disease in human populations.

Effects on evolutionary biology. The rapid growth rate and complex life cycles of many infectious diseases agents facilitate the evolution and emergence of new pathogens through horizontal gene transfers or cross-species transmission of microbes. For example, major epidemic influenza virus strains are known to arise by mixing and genetic reassortment of genes from influenza viruses of humans, pigs, and migratory waterfowl (Webster et al., 1992). In recent years there has been growing interest in understanding how disease agents evolve to enhance their own survival, in particular by developing resistance to antibiotics and other treatment drugs (e.g., Ewald, 1994). Climate and ecological changes could plausibly influence the evolution and adaptation of disease pathogens. Thus far, however, there has been very little research to explore such questions, and thus almost nothing is known about potential impacts.

Indirect effects operating through ecological changes. Weather and climate changes can lead to epidemic disease by altering local ecosystems. One example of this is Rift Valley fever, an epidemic viral illness in eastern Africa. Heavy rainfall with ground saturation and pooling of water in surface depressions ("dambos") leads to increased hatching of the mosquito vectors of the virus and subsequently increases viral transmission (Linthicum et al., 1988). Another example is onchocerciasis (or river blindness) caused by a parasite common in tropical West Africa and South America, where the breeding of the disease vector, the Simulid fly, is governed by river water flow (WHO, 1985). The ecosystem instabilities brought about by climatic changes can give rise to new interactions among hosts and infectious disease agents, possibly accelerating the problem of emerging infectious diseases.

Indirect effects operating through changes in human activities. Changes in weather often bring about changes in human activities that can influence infectious disease transmission rates. For example, warm ambient temperatures often lead to greater use of central air-conditioning systems, which in turn may harbor and spread the bacillus responsible for legionnaires' disease (Garbe et al., 1985). Another example is the fact that school holidays often bring about a decline in influenza and other respiratory virus illnesses simply by reducing crowding and contacts. In the tropics, drought can lead to an increase in dengue because more people store water in open containers, thus increasing the number of breeding sites for mosquitos (Moore et al., 1978).

Ultimately, these climatic impacts must be placed in the context of all the other factors that influence infectious disease dynamics (see Figure 3-5), including land-use changes, transportation and migration patterns, urban crowding, widespread use of antibiotics, and changes in public health infrastructure. (These factors are discussed in depth in the following section.) The disease impacts of climatic changes can in some cases be dampened through natural immune re-

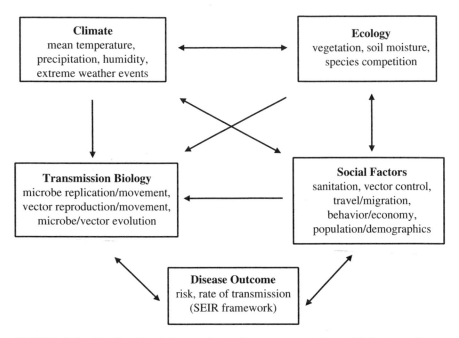

FIGURE 3-5   The "web" of factors that influence transmission of infectious disease agents.

sponses at both the individual level and the population level ("herd immunity"). There are also numerous types of social responses that can modulate the impact of epidemics. Fleeing from urban epidemic epicenters and quarantine of travelers are centuries-old responses. Modern responses include the use of drugs and vaccines to interrupt transmission and/or provide immunity.

Over the long term, a wide array of other largely unforeseeable changes could greatly affect the type and global distribution of infectious diseases. For instance, it is possible that 50 years hence the potential to control diseases such as malaria and dengue could be greatly strengthened by vector eradication programs, improved sanitation/public health systems, or the development of drugs or vaccines. Likewise, it is possible that in the coming decades diseases that are now very rare will become more prevalent or that new strains of infectious agents will emerge. All of these uncertainties greatly complicate any attempts to assess future disease threats.

Extreme Weather Events and Infectious Disease Outbreaks

Extreme weather events such as floods, droughts, and hurricanes occur as isolated incidents but can also be manifestations of longer timescale climate

---

Box 3-4
Progression of Climate and Health Assessments

Often in the study of a complex environmental issue, as scientific understanding of the issue progresses, investigators increasingly come to appreciate the broader context in which the issue is embedded. The study of climate and disease linkages has followed this typical progression, and this is reflected in successive assessment reports from the Intergovernmental Panel on Climate Change (IPCC). The IPCC's first assessment report in 1990 contained only a couple of very general statements that "parasitic and viral diseases . . . have the potential for increase and reintroduction in many countries" due to climate change. Subsequent IPCC reports issued in 1995 and 1998 included more detailed discussions about climate influences on specific diseases, placed in different regional contexts. These analyses considered the difference between direct and indirect effects and emphasized the role of socioeconomic factors in determining disease distribution. The third IPCC assessment report (IPCC, 2001) elaborates further on some of these issues and explicitly addresses the methodological difficulties of attributing causation between climate and other coexistent factors. This increasing number of caveats and uncertainties may at first appear to indicate a lack of progress in scientific understanding of climate/disease linkages. In reality, though, the opposite is true; it reflects a growing understanding of the complexity of the issue and of the need for integrated research approaches.

---

variations. For instance, El Niño events typically increase the frequency of tropical storms and hurricanes in the eastern Pacific, and La Niña events favor Atlantic hurricane formation (Saunders et al., 2000). Long-term climate change may increase the risks of extreme events such as floods and droughts occurring in particular regions and could possibly affect the frequency and intensity of mid-latitude storms and hurricanes; however, such projections are highly uncertain at this time.

While the primary risks associated with extreme weather events are injury and death that result directly from the event itself, there is also concern that weather-related disasters can lead to infectious disease outbreaks. The risk factors associated with disasters include (as discussed in Noji, 1997):

- population displacement and crowding in emergency shelters, which increases the probability of disease communication, particularly when sanitary facilities are insufficient;
- migration due to drought or other disasters, which exposes refugees to locally endemic disease against which they have no immunity;
- disruption of healthcare services due to physical damage to roads, hospitals, etc.;
- malnutrition, which makes people more susceptible to some diseases;

- unavailability of safe drinking water due to contamination of wells and reservoirs from overflow of sewage or failure of water treatment facilities;
- increase in vectors due to a greater number of breeding sites or disruption of vector control programs;
- greater exposure to vectors due to lack of shelter.

A common myth associated with natural disasters is that disease epidemics are inevitable. It must be recognized, however, that if disease-causing pathogens are not present in an affected area and are not introduced into the area after a disaster, then disease will not occur even if environmental conditions are ideal for transmission. This may explain why there have been almost no documented disease outbreaks associated with a natural disaster in the United States in over three decades. One possible exception is an outbreak of coccidioidomycosis in California in the wake of a major windstorm (Flynn et al., 1979). It is important to characterize accurately the disease risks associated with different types of extreme weather events, so that limited resources are not needlessly diverted from other, perhaps more critical, facets of a disaster relief effort.

The risks associated with waterborne disease are fairly well understood, and there are numerous examples of enteric disease outbreaks following floods, largely due to disruption in public health and sanitation services (i.e., water treatment). With vector-borne diseases, however, the picture is much less clear. Hurricanes and floods can affect vector populations in numerous and complex ways that can either increase or decrease disease transmission rates. For instance, mosquitos and other vectors can be physically washed away by strong winds and waterflows, which can temporarily suppress disease risk; but as the waters recede and leave behind pools of standing water, the vector populations may rebound dramatically with the greater availability of breeding sites. If the landscape is significantly altered by a disaster, this can have long-term effects on the vector ecology of a region.

There are few studies documenting infectious disease outbreaks associated with disasters, because collection of epidemiological data is not a top priority concern during a crisis situation. Bissell (1983) points out that many epidemiological studies of disasters cover only the first several weeks after an event and thus do not account for the significant lag time in the development of some epidemics (e.g., hepatitis A). Some diseases have a long incubation period and can take several months before reaching their maximum distribution.

While most studies focus on the effects of hurricanes and floods, droughts also have the potential to affect infectious disease risk. Droughts generally develop over a longer timescale than other natural disasters and thus offer more opportunity for interventions. Famine is the most direct risk associated with drought, but the resulting malnutrition can also increase risk of severe illness among infected individuals (Tompkins, 1986). There have been a few instances of cholera outbreaks associated with drought conditions, due to disruption of clean water supplies. Also, drought could increase the risk of meningoccocal meningitis in regions of Africa and Asia where the disease is associated with dry conditions and dust storms (Greenwood et al., 1984).

OTHER FACTORS THAT AFFECT INFECTIOUS DISEASE DYNAMICS

The primary focus of this report is on the relationships between climate and infectious diseases, but in order to understand this relationship it is also necessary to appreciate the many other factors that contribute to the distribution and dynamics of infectious diseases. These include factors such as land-use patterns; social, demographic, and geographical considerations; transportation and migration patterns; and public health interventions. Some of these factors are closely interrelated and are themselves influenced by climate, often in indirect and subtle ways. The complexity of these relationships gives credence to the need to appreciate the "web of causation" in public health.

As described by May (1958), for disease to occur, populations must behave in ways that allow the mutual contact of agents and hosts. Constraints on the geographical distribution of disease agents include landscape characteristics, proximity to breeding areas, and proximity to hosts themselves; while constraints on host behavior may include habitat, cultural influences, tradition, and transportation availability. These concepts, which are explored further in the following sections, are essential to understanding the ecology of disease.

Land Cover / Land Use

Land-use and land-cover changes can exert a profound influence on the distribution of infectious diseases. For instance, malaria vectors tend to be concentrated on land that is exposed to sunlight and that contains pools of water, thus land-cover changes that impact the amount of sunlight in vector-breeding sites can alter the dynamics of malaria transmission (Walsh et al., 1993). The snails that serve as intermediate hosts for schistosomiasis require immobile or slowly moving water. Rice paddies in China, slow-moving water such as the Nile upriver of the Aswan dam, and inland pools of water are all environments in which highly endemic schistosomiasis is frequently found, and modification of these environments has been important in schistosomiasis control efforts (Hairston, 1973).

The proliferation of Lyme disease in the United States has largely been the result of changing land-use patterns. Following European settlement of the northeastern United States, much of the forested land was cleared for agricultural use. As agricultural activity subsequently moved westward to allow for the growth of suburban and new semi-urban settlements, land on the periphery of towns was replaced by second-growth forest. This created an ideal habitat for the white-tailed deer, hosts of the ticks that transmit the disease. In recent years, population pressures and the demand for suburban housing located on the edge of forests have led to more human contact with deer habitat and thus with Lyme disease vectors (Mayer, 2000).

Transportation and Migration

The spread of infectious disease agents is greatly affected by human travel patterns. Migration is one of the means by which diseases spread, either because migrants bring new pathogens with them to their destinations or because the migrants themselves constitute susceptible populations and lack immunity to endemic diseases in their areas of settlement. This is true for both forced migration caused by political/religious oppression and natural disasters and for voluntary migration of people seeking new social or economic opportunities. For instance, migration has been linked to disease transmission patterns of malaria in Africa (Prothero, 1965), and there are examples of outbreaks of malaria among refugees with low levels of acquired malaria immunity in Thailand, Pakistan, Sudan, and Nepal.

Modern transportation has been implicated in the spread of disease pathogens and vectors. Jet transportation in particular makes it possible for an infectious disease to spread rapidly from area to area within a continent or from one continent to another. This has been the case with influenza, where it appears that new strains initially spread from Southeast Asia to other areas of the world. Infected people, who may be asymptomatic, can infect fellow passengers and susceptible people at their destinations. This is particularly the case for diseases with easy transmissibility such as tuberculosis (Kenyon et al., 1996) and other respiratory diseases.

Transportation has also been the putative means by which non-respiratory infectious diseases may be introduced into new areas. For example, antibiotic-resistant gonococci (which cause gonorrhea) initially were found in Asia and then spread to the United States (Gordon et al., 1996; Knapp et al., 1997), clearly following human sexual contact patterns. The movement of HIV/AIDS in Africa has been linked to the sex trade among those engaged in long-distance trucking in East Africa (Pickering et al., 1997).

The development of more highly interconnected areas as regions grow economically is a long-term factor in the transmission of infectious diseases. For instance, the spread of cholera through the United States in the nineteenth century directly reflected the changing connectivity of the urban system. In the earlier cholera epidemics of the century, the spread was based on distance from the portals of introduction of cholera into the country. As individual urban areas became more interconnected through the development of railroads and road systems, the disease spread rapidly from one major metropolitan area to another and to smaller towns within range of these cities (Pyle, 1969).

Transportation vehicles themselves can contribute to the spread of vectors to new areas. The concept of "airport malaria" arose from numerous reports of limited malaria outbreaks among populations surrounding airports in temperate non-endemic areas such as the United States, England, and Northern Europe. The clustering of cases around international airports and the subsequent experimental confirmation that anopheline mosquitos could survive a long-distance flight in the wheel wells of jet aircraft demonstrates the potential for air transportation to facilitate the spread of disease vectors (e.g., Giacomini et al., 1997; Guillet et al., 1998). Similarly, one of the Asian vectors of dengue, the mosquito *Aedes albopictus,* is thought to have been transported to Houston in wet tires being carried aboard a container ship (Moore and Mitchell, 1997). This mosquito may be capable of transmitting dengue in the Americas as it spreads throughout the region.

Social and Demographic Patterns

Social and demographic patterns are major influences on the distribution and spread of infectious disease; and infectious diseases, in turn, exert a major influence on population structure. For example, the "epidemiological transition" is a commonly observed pattern in the development of most societies, wherein most mortality is initially due to infectious causes, but with improved sanitation and nutrition, mortality due to infectious diseases decreases while mortality due to non-infectious causes such as heart disease, stroke, cancers, and accidents increases. Along with this transition comes an increase in life expectancy, and the age structure of the population thus shifts to one where there are proportionately more middle-aged and older individuals.

Population density is another important factor to consider. Contagion is facilitated by population concentration because infected individuals have a higher probability of contact with susceptible members of the population. Population density has been linked with increasing ease of transmission of airborne infections, waterborne diseases, and sexually transmitted infections. Rapid urbanization and population growth in Southeast Asia, concurrent with spread of the urban mosquito vector, are considered to be primary reasons why dengue has become endemic in the region in recent years.

Poverty exerts profound effects on infectious disease patterns, although it is sometimes difficult to separate the effects of population density from other confounding effects of poverty. For instance, immunity can be compromised by malnutrition, HIV infection or other concurrent disease, or addiction to drugs or alcohol (Farmer, 1999). The typical pattern in the developed world is a disproportionate concentration of violence, poverty, and drug addiction in inner-city neighborhoods (Friedman et al., 1996). These neighborhoods have multiple characteristics that are conducive to higher prevalence and easier transmission of infectious diseases, including high rates of homelessness, co-infection and co-morbidity, and poor sanitation and nutrition. Sometimes hospitals and clinics can even provide a locus for spread of an infection to staff and other patients through inadequately sterilized equipment and inadequate isolation measures.

Human contact with previously uninhabited regions such as tropical rainforests has increased due to activities such as deforestation and land clearance schemes such as commercial rubber plantations in Malaysia. This enhances the possibility of human contact with infectious disease cycles in primates and other animals and therefore creates a greater risk of transmission of zoonotic agents to humans. Since the populations involved in these activities are usually very mobile, the potential for rapid transportation of pathogens out of their usual ecological niches is high. This is one reason why there is great concern over the transportation of virulent tropical hemorrhagic fevers such as Ebola, Marburg, and Lassa fevers from their endemic areas to more settled regions in both Africa and other continents. In tropical areas where previously endemic infectious diseases have remained highly localized, greater global interconnectivity is being reflected in the redistribution and spread of these diseases.

Household design and architecture can also influence patterns of vectored disease transmission. For example, the ecology of Chagas disease (American trypanosomiasis) has been transformed as the reduviid or "kissing" bugs that spread this highly prevalent disease have adapted to live in certain types of housing (Bastien, 1998). Second, perpetuation of the cycle of urban dengue fever depends on the proximity of pools of water in tires, cisterns, and on the roofs of houses. Third, some anophelines mosquitos find conducive habitats in urban and rural housing structures, which puts the residents of these houses at greater risk of contracting malaria.

Water development projects have been clearly linked with the intensification of certain disease cycles. Following completion of the Aswan Dam, upriver schistosomiasis increased in prevalence and shifted from a form that primarily affected young children to one that primarily affected young adults in the labor force (Mobarak, 1982). Rift Valley fever and malaria also were linked to the creation of lakes along the Nile and the increased length of the shoreline.

Health Services and Public Health Intervention

The availability of public health services can significantly affect the distribution of disease, since the very purpose of such services is to stem the spread of disease in populations. Vaccination is a specific intervention aimed at preventing the occurrence of disease in individuals and in so doing can reduce the incidence of an infectious disease in populations. Smallpox was the first infectious disease to be eliminated through a program of targeted intervention and vaccination. Polio is now relatively close to being eradicated through vaccination programs, and in many areas childhood diseases such as measles and chickenpox have been well controlled. This has all been due to public health programs mandated by state or national laws and to the efforts of international public health agencies such as the World Health Organization and public service groups such as Rotary International.

The development of antimicrobial agents has also profoundly altered patterns of infectious disease. For instance, the widespread availability of penicillin made it possible to intervene in the progression of streptococcal pharyngitis to rheumatic fever; the development of anti-tubercular agents reduced the prevalence of tuberculosis; sexually transmitted infections such as syphilis, gonorrhea, and chlamydia are highly treatable with antimicrobials; and community-acquired bacterial pneumonias are similarly treatable. However, drug resistance and antimicrobial resistance are increasing problems with some of these diseases.

Public health measures such as improved sanitation, better nutrition, and environmental modification have been of great significance in altering patterns of disease transmission. Indeed, some diseases have been eliminated in specific regions because of these measures. For example, endemic malaria was eliminated in Trinidad through programs for pesticide application, draining swamps, and public education about reducing exposure to mosquitos. Also, the elimination of schistosomiasis in most Caribbean nations, Israel, and elsewhere has been a consequence of eliminating water pools, altering subsurface flora and water flow characteristics, and the application of molluscicides to eliminate the snails that are responsible for transmission of the disease.

Box 3-5
Dengue on the U.S./Mexico Border

A striking example of the influence of non-climatic factors on disease distribution is the occurrence of dengue in regions that lie along the U.S./Mexico border. Between 1980 and 1996, there were 50,333 confirmed cases of dengue recorded in three Mexican states that border the Rio Grande, while there were less than 100 cases on the other side of the river in Texas, and most of these cases were among individuals who had recently visited Mexico (Texas Department of Health, personal communication).

The climatic differences between these two regions are insignificant, and surveillance data indicate that mosquito density is actually higher in Brownsville, Texas, than in Reynosa, Mexico (3.8 vs. 2.7 *Aedes aegypti* pupae per person, respectively). Thus, in this case the dramatic discrepancy in disease occurrence can be attributed to differences in living conditions. The proportion of homes with air conditioning is higher in Brownsville than Reynosa (42 vs. 2 percent) and the proportion of homes with adequate window screens is also higher in the U.S. community (57 vs. 15 percent). Another important factor enhancing transmission of disease agents is the greater proportion of homes in Reynosa with adults and infants present during the day, the biting period of the vector. These factors result in significant differences in human exposure to mosquito bites and transmission.

Water treatment is a critical measure for protecting public health. Major advances in the control of waterborne diseases began about 1900, when diseases such as cholera and typhoid were dramatically reduced in the United States through the disinfection of drinking water with chlorine (Baker, 1948). By the 1940s and 1950s, sewers and primary treatment of wastes emerged, and secondary treatment came about in the 1970s along with disinfection of wastewater. Despite these advances, waterborne disease continues to occur in many places and affects food supplies through the contamination of irrigation waters and fishing and shellfish harvesting areas. Some of the reasons for this include aging infrastructure of sewage and water treatment facilities in the face of increasing water demands; inability to identify the microorganisms associated with many diseases; newly recognized microorganisms that are more resistant to disinfection (e.g., *Legionella, Cryptosporidium, Giardia*); contamination of water supplies with sewage overflow or concentrated fecal sources from animal farming operations; and increasing non-point sources of pollution, including septic tanks.

4

Climate Influences on Specific Diseases

The previous chapter reviews the general ways that the emergence and transmission of disease agents can be influenced by climate. The specific mechanisms underlying these linkages vary widely from one disease to another, as does our understanding of these linkages. This chapter focuses on a few specific diseases, chosen because they offer the opportunity to explore a diverse range of vectors, transmission pathways, geographic regions, and relationships to climate. In each case, the disease's impacts and global prevalence are described, as well as the life cycle of the pathogen and vector and the ways that this life cycle can be influenced by climate and other factors. Table 4-1 summarizes the different categories of diseases that are discussed.

DENGUE

Disease description. In terms of the number of human infections occurring globally, dengue is considered to be the most important arthropod-borne viral disease in humans. The symptoms of dengue include fever, severe headache, muscle and bone pain, and occasionally shock and fatal hemorrhage. The average case fatality ratio for dengue hemorrhagic fever is about 5 percent. Dengue is caused by four distinct flavivirus serotypes, and there is only short-lived cross-immunity from each serotype. The variability in the types of virus circulating and the nature of dengue antibodies in the human population interact to alter the frequency of infection and cases of dengue (Gubler, 1988).

There is a global pandemic of dengue, distributed throughout the tropics, and it is estimated that there are roughly 50 million cases of dengue infection

45

TABLE 4-1 Examples of Pathogen Transmission Associated with Environmental Factors

Transmission Category	Examples of Pathogens	Routes/Sources of Transmission	Environmental Factors	Other Factors
Vectorborne	Malaria Dengue	Routes/Sources of (anthroponotic, indirect)	Rainfall, temp., and relative humidity influencing vector and/or pathogen levels/distribution	Vector control Human migration or travel
	Lyme disease St. Louis encephalitis Rift Valley fever	Insect/tick-animal reservoir-humans		
	Hantavirus	Rodent-humans (zoonotic, indirect)		
Airborne	Influenza	Respiratory (anthroponotic, direct)	Relative humidity and pathogen survival	Level of endemic infection and crowding
Water- and food-borne	Cryptosporidium	Fecal-oral (humans and animals) (zoonotic, direct)	Rainfall and water contamination	Management of wastes and water treatment
	Vibrios	Naturally occurring (anthroponotic, direct or indirect)	Temp. and salinity influencing growth	Algal blooms, uncooked shellfish

worldwide very year (WHO, 1998a). Dengue virus is carried by *Aedes aegypti*, a mosquito that preferentially feeds on humans and is particularly suited to urban environments. When a mosquito bites an infected person, it acquires the virus through the blood meal. In the insect's gut the virus replicates and spreads to the salivary gland, priming the mosquito for transmission of infection to another person.

Climate influences. The abundance of dengue vectors depends in part on the availability of breeding sites, primarily containers such as drums, discarded tires, and leaf axils that are filled with water either manually or by rainfall. Where containers are manually filled for water storage, vector abundance is largely independent of rainfall, as seen in some areas of Bangkok, Thailand (Sheppard et al., 1969). In contrast, in southwestern Puerto Rico a high proportion of available breeding sites are discarded containers such as tires, bottles, and tin cans that become filled with rainfall. In this case there is a clear correlation between rainfall and *A. aegypti* abundance (Moore et al., 1978). Heavy rainfall tends to overflow containers and can thus discourage vector breeding, while extended drought conditions have in some cases led to higher vector abundance due to an increased use of water storage containers.

Saturation deficit, a parameter taking into account both temperature and relative humidity, affects the survival of eggs and adults. Newly laid eggs are subject to desiccation, and adults can experience moisture-related reductions in survival throughout their lifetimes. However, saturation deficits high enough to reduce egg or adult survival, according to the few published studies, are rarely encountered in humid tropical locations (Southwood et al., 1972). Atmospheric moisture also influences the rate of water loss from containers, which in turn affects vector abundance.

Temperature affects the potential spread of the virus through each stage in the life cycle of the mosquito. Adult and immature *A. aegypti* survive in a broad range of temperatures, from about 5°C to 42°C, although temperatures below 20°C reduce or prevent eggs from hatching. Temperature also influences the time required for embryonic, larval, and pupal development and plays a major role in the frequency of biting. Temperature also affects the extrinsic incubation period (EIP), the period between when the mosquito imbibes virus-laden blood and actually becomes infectious. At lower temperatures the EIP is longer and the mosquito is less likely to survive long enough to transmit the virus. In the *A. aegypti*/dengue system, EIP is a non-linear function of temperature such that even small changes in temperature introduce a seasonality into transmission dynamics (Focks et al., 1995).

Recent studies of the distribution and epidemiology of dengue viruses suggest that projected climate warming is generally expected to increase the intensity of transmission. For instance, Patz et al. (1998) estimate that for regions where dengue is already present, a mean temperature increase of about 1°C increases the aggregate epidemic risk by an average of 31 to 47 percent. Higher infection rates translate into a greater number of individuals who have experienced multiple infections and thereby may have an elevated risk for developing serious dengue illness (Halstead, 1988; Jetten and Focks, 1997). Another possibility to consider, however, is that climate change may lead to lower average atmospheric moisture levels in some regions, which could actually reduce dengue transmission.

MALARIA

Disease description. Malaria is one of the most common vector-borne diseases in the tropics. The pathogens are protozoan parasites in the genus *Plasmodium* that are transmitted to humans by the *Anopheles* mosquito. In humans the malarial parasites infect red blood cells, causing periodic chills and fever, and in some species of *Plasmodium* the parasites can ultimately cause death. Insufficient vigilance in maintaining mosquito control measures, development of insecticide resistance by the vectors, and the emergence of drug-resistant strains of malaria have contributed to the resurgence of this disease throughout the tropics. Worldwide prevalence of the disease is estimated to be in the order of 300-500 million clinical cases each year, causing over 1 million deaths annually. About 90 percent of these cases occur in sub-Saharan Africa, and most of the others are found in other tropical regions (WHO, 1998b).

Humans get malaria from bites of infected female mosquitos. The parasites replicate asexually in the red blood cells of the human host and are responsible for the clinical manifestations of disease. Some of the parasites differentiate into sexual stages that continue the transmission cycle when ingested by another mosquito during a human blood meal. Sexual reproduction of the parasite occurs in the mosquito.

Climate influences. Temperature, rainfall, humidity, and wind each play a role in determining the distribution and incidence of malaria. These factors govern the distribution, prevalence, rate of development, life span, and feeding frequency of the *Anopheles* mosquitos. Temperature also plays a fundamental role in the rate of multiplication of the parasite in the mosquito. Although each species of *Anopheles* has a different ecology, as a general rule, warmer temperatures mean that the mosquito develops more rapidly and feeds more frequently and earlier in its life cycle and that the parasite within the mosquito develops and multiplies more rapidly. Thus, malaria tends to occur less frequently at higher altitudes and latitudes, at least in part because these regions are associated with colder temperatures.

Several studies indicate that malaria has spread or increased in transmission intensity following temperature and/or rainfall variations brought about by El Niño events. These epidemics involve unstable malaria transmission, often in desert and highland fringes. In northern Pakistan, higher temperatures associated with El Niño correlated with increased malaria incidence (Bouma et al., 1994). Malaria outbreaks in the former Punjab province of India were more likely in the year after an El Niño with a relative risk of 4.5, and in El Niño years in Sri Lanka with a relative risk of 3.6 (Bouma and van der Kaay, 1996). There was a highly consistent association between El Niño years and outbreaks of malaria in Venezuela (Bouma and Dye, 1997) and Colombia (Bouma et al., 1997). Linblade et al. (1999) found that in the highland region of southwestern Uganda during the El Niño event of 1997-1998, an increase in rainfall was positively correlated with an increase in vector density and incidence of malaria (almost three times the mean incidence of the preceding five years). These associations are strong and consistent and imply that ENSO forecasts may have value in predicting disease risk in particular regions, even though the mechanisms underlying these associations are not always apparent.

At present, there is insufficient evidence to clearly attribute an increase in malaria incidence or its geographic spread to long-term global warming patterns. While some studies have predicted that global climate change could potentially lead to widespread increases in malaria transmission by expanding mosquito habitat range and "vectorial capacity" (Martens et al., 1999), other models have projected only a negligible extension in potential malaria transmission due to climate change (Rogers and Randolph, 2000; these studies are discussed further in Chapter 5). Regardless, these models can only simulate the potential transmission risk; actual occurrence of the disease is determined largely by socioeconomic factors. For instance, historically malaria was common in temperate regions but is no longer found in such regions because of environmental sanitation, mosquito control (e.g., draining swamps that breed some species of *Anopheles*), and lifestyle changes that allow people to minimize exposure to mosquitos (e.g., by screening houses and using air conditioning). Similarly, Reiter (2000) emphasizes that an increase in the potential malaria burden due to future warming trends could be offset by other factors, especially human interventions, as documented by the fact that during the warming trend of the past three centuries, the geographic distribution of malaria diminished rather than increased.

ST. LOUIS ENCEPHALITIS

Disease description. St. Louis encephalitis is an acute inflammation of the brain caused by a mosquito-transmitted virus in the family *Flaviviridae*. Mild infections cause fever and headache, but more severe infections may cause encephalitis, with symptoms of headache, high fever, neck stiffness, stupor, disorientation, coma, and paralysis. The disease occurs in North America, and the virus is closely related to the Japanese encephalitis virus of Asia and West Nile virus of Africa, Eurasia, Australia, and North America. The original epidemic in St. Louis County, Missouri in 1933 was associated with an extremely dry summer, during which the city's open sewage ditches did not flush and mosquito breeding was unusually abundant (Kinsella, 1935). There have been 4,478 reported human cases of St. Louis encephalitis in the United States since 1964, with an average of 128 cases reported annually (Shope and Tsai, 1998). One in 338 infected people becomes sick, and the case-fatality ratio is between 2 percent in young people and 23 percent in persons over 70 (Monath, 1980).

The urban mosquito vector of St. Louis encephalitis is *Culex pipiens*, which maintains the virus in a life cycle involving several species of wild birds. The mosquito is infected when it takes a blood meal from a viremic bird. The virus multiplies in the mosquito, and after an extrinsic incubation of about two weeks, the virus appears in the saliva and can be transmitted either to another bird or to a person. The virus also has a rural cycle in the western United States, where the *Culex tarsalis* mosquito is the major vector. This mosquito breeds in flood waters and is abundant in irrigated fields and riverine flood plains, and thus excess rainfall and abundant snowmelt can favor its breeding (Monath, 1980).

Climate influences. Temperature has a major influence on the development of the *Culex* mosquito and St. Louis encephalitis virus in the mosquito. Epidemics occur primarily in southern latitudes where the mean June temperature is 21°C or above. Monath and Tsai (1987) compared retrospectively 15 urban U.S. outbreaks of St. Louis encephalitis, and found statistically significant associations with increased precipitation and temperature in January, decreased temperature in April, and increased temperature in May. Seven of 15 epidemic years had all of these characteristics, as contrasted to only two of the 130 non-epidemic years.

Reeves et al. (1994) studied the ecology of St. Louis encephalitis in mosquitos of California by comparing the warm southern San Joachim Valley with the colder northern Sacramento Valley. Based on their findings about the dynamics of mosquito infection and life cycles, they predicted that if a 5°C warming were to occur, St. Louis encephalitis would become less prevalent in the warmer south and would be distributed farther north in California, Oregon, and Washington. This prediction, however, did not take into account socioeconomic factors that might modify exposure of persons to infected mosquitos.

RIFT VALLEY FEVER

Disease description. Rift Valley fever (RVF), first described in 1930 during an epizootic among domestic livestock in the Rift Valley of Kenya, is a potentially fatal disease of humans that results from infection with RVF virus (Daubney et al., 1931). During the past half-century, dozens of animal and human outbreaks of RVF have occurred, mostly in sub-Saharan Africa (Meegan and Bailey, 1988). Transport of infected mosquitos or animals into Egypt resulted in outbreaks during 1977-1978 and again in 1993-1994 along the Nile River.

Symptoms of RVF range from unnoticeable or mild disease in adult sheep and cattle to abortion among pregnant ungulates and fatalities in young animals (Shimshony and Barzilai, 1983). Human disease may be mild but can involve abrupt onset of high fever, severe headache, myalgia, and incapacitation for several days (Peters and Meegan, 1981). Some patients develop fatal hemorrhagic fever or encephalitis. There is a 1 percent case fatality rate in humans. Low-level endemic transmission occurs regularly throughout much of the African continent, but most of this remains unrecognized due to inadequate surveillance and health care facilities (Meegan and Bailey, 1988).

RVF virus is transmitted by mosquitos to both domestic animals and people. Although more than 30 mosquito species are potential vectors, certain floodwater *Aedes* species that emerge from temporary ground pools following seasonal rains are recognized as the important enzootic vectors (McIntosh and Jupp, 1981). Epizootic transmission often occurs during periods of heavy rainfall in East Africa (Davies et al., 1985) when particular *Culex* species become infected. Transmission of RVF virus depends on complex interactions among mosquito vectors, non-human vertebrates, and human hosts, all of which are linked to various environmental factors (Wilson, 1994). Mosquitos perpetuate transmission either by direct infection of eggs or by acquiring infection during a blood meal and transmitting it during a subsequent feeding. Not only is species-specific mosquito abundance important, but densities of certain vertebrates, especially domestic ungulates, are crucial factors in zoonotic outbreaks. Such outbreaks induce protective immunity in all infected animals, thus reducing the number of susceptible animals after an epizootic. Human proximity to infected animals and mosquitos partly determines disease risk to people. *Aedes* mosquitos that infect animals rarely bite humans, so abundance of other mosquitos (notably *Culex* species) may be important for human disease.

Climate influences. Studies of climate variability and RVF activity have focused on precipitation and epizootics. Periods of excessive rainfall are believed to increase the egg hatching and larval survival of certain African *Aedes* floodwater mosquito species. Extensive transmission to humans, however, would seem to require the buildup of *Culex* populations. Recently, Linthicum et al. (1999) analyzed the historical pattern of eight recognized RVF outbreaks in East Africa since 1950 and claimed that each followed periods of abnormally high rainfall. Using Pacific and Indian ocean sea surface temperature anomalies and satellite-based vegetation indices, these authors suggest that outbreaks may be predicted up to five months in advance of their occurrence. Whether this association is strong and consistent enough to allow such forecasts remains to be seen.

HANTAVIRUS

Disease description. In 1993 an outbreak of acute respiratory distress with a high fatality rate (>50 percent) occurred in the population of the Four Corners area of New Mexico, Arizona, Colorado, and Utah. The cluster of cases was caused by a newly recognized virus that was given the name Sin Nombre virus, a hantavirus (Nichol et al., 1993). The disease, hantavirus pulmonary syndrome (HPS), was soon recognized over a wide area of North and South America.

Patients with HPS typically exhibit fever and muscle aches lasting three to five days; shortness of breath and coughing symptoms then develop rapidly, requiring hospitalization and ventilation. Since first identified in May 1993, 260 cases of HPS have been reported in the United States, with the peak, 80 cases, occurring in 1993. The Sin Nombre virus is a rodent-borne pathogen belonging to the bunyavirus family of RNA viruses. Deer mice (*Peromyscus maniculatus*) are the most common reservoir in the southwestern United States, and they shed virus in their urine, droppings, and saliva (Childs et al., 1994). The virus is mainly transmitted to people when they breathe in contaminated air. Molecular tests of archived postmortem tissues from humans and rodents revealed that HPS was not a new disease, but that it was only newly recognized as a result of this unusual cluster of cases in 1993.

Climate influences. Data from an ongoing study of *Peromyscus* rodents revealed that in 1993 populations of this animal in some parts of New Mexico were as much as 10 times greater than average (Parmenter et al., 1993). The hypothesis was advanced that the unusually heavy rainfalls occurring during the 1992-1993 El Niño led to an abundant food supply for rodents, followed by a rodent population explosion. Deer mice readily entered homes and farm buildings, which resulted in greater exposure of humans to hantavirus-infected rodent excreta.

To test this hypothesis, investigators used satellite imagery and precipitation estimates to identify environmental conditions that were associated with the sites where HPS cases occurred (Glass et al., 2000). The clustering of cases was found to be associated with areas of heavy vegetation, but the data failed to fully support the connections between El Niño and HPS risk as proposed above. Also, at this point such associations may be confounded by the fact that more people have begun taking actions (such as rodent-proofing their houses) to mitigate risk of exposure to the virus.

LYME DISEASE

Disease description. Lyme disease is caused by a spirochete bacterium (*Borrelia burgdorferi*) that is transmitted by tick bites. The disease typically presents with a characteristic "bull's-eye" rash, accompanied by symptoms such as fever, fatigue, headache, and muscle and joint aches; although some infected individuals show no signs of illness. It is rarely, if ever, a primary cause of death. Lyme disease occurs throughout many areas of the northeastern United States and parts of the northern Midwest (especially Wisconsin and Minnesota) and is found at lower frequencies in the southeastern states and on parts of the West Coast, especially northwestern California. Lyme disease is also found in parts of Europe and eastward across Asia (Dennis, 1998). The annual number of cases in the United States per year has increased about 25-fold since national surveillance for the disease began in 1982. A mean of approximately 12,500 cases was reported annually to the Centers for Disease Control and Prevention from 1993 to 1997.

In all areas where Lyme disease occurs, ticks of the genus *Ixodes* appear to be the vector responsible for most transmission. The predominant vector in the eastern United States is the deer tick (*Ixodes scapularis*) and in the western United States the western black-legged tick (*Ixodes pacificus*). The tick has three motile stages (larvae, nymphs, adults), and each of these stages must find, attach to, and feed on the blood of a vertebrate to eventually molt to the next stage or, in the case of adult females, to reproduce. Mated and blood-fed female ticks lay perhaps 2,000 to 3,000 eggs, which hatch into larvae and begin another generation. Adult ticks in North America feed predominantly on large vertebrates, particularly white-tailed deer (*Odocoileus virginianus*). Abundance of deer has been linked to abundance of ticks in infested areas. The principal reservoir of the spirochete, however, is the white-footed mouse (*Peromyscus leucopus*). Larval or nymphal ticks become infected while feeding on an infected mouse and are then capable of transmitting infection to another vertebrate during a subsequent blood meal. Occasionally, a tick will bite and feed on a human, sometimes resulting in Lyme disease (Dennis, 1998; Wilson, 1998).

Climate Influences. Many factors appear to influence the spatial and temporal patterns of Lyme disease risk. Transmission among natural hosts and humans is determined by local abundance and survival of ticks, the percentage of ticks that are infected, abundance of hosts, human activity in tick habitats, and people's knowledge and awareness of Lyme disease prevention. A few studies have demonstrated how micro-climate influences the survival and activity of deer ticks (e.g., Bertrand and Wilson, 1996), suggesting that climate patterns may be important in determining Lyme disease risk; but research thus far has not yet fully elucidated the contributions of climate and other factors to transmission dynamics.

Climate change scenarios often forecast that some regions of the United States may become warmer and moister, leading to speculation that the range of deer ticks carrying Lyme disease might expand. However, the current distribution of the deer tick and of Lyme disease in the Unites States spans a wide range of climatic conditions, and deer ticks are already abundant in parts of the country where cold extremes are common. Extremely cold regions of North America and Northern Europe could perhaps support survival of local *Ixodes* ticks if climate warming were to occur, but whether this would translate to new cases of Lyme disease is highly speculative. If the range of ticks were to expand, this could set the stage for expansion of disease, but many other factors that contribute to human risk would also have to develop (e.g., competent reservoirs, human behavior, seasonal activity patterns).

Although climate is important, various other factors appear to be primarily responsible for risk of this vector-borne disease. The range of factors that presently limit distribution of the vector remain poorly understood, but research suggests that microclimate, abundant hosts, and suitable vegetation and soil habitat are important. The tick has not yet become established or widespread in apparently appropriate environments in many areas of the United States, thus indicating that even where suitable microclimate, host, pathogen, and human contact conditions appear to exist, Lyme disease may not be present.

INFLUENZA

Disease description. Influenza or "flu" is a viral infection of the respiratory tract that affects millions of people globally every year. Influenza is highly contagious and can cause severe complications such as pneumonia, particularly in children, the elderly, and other vulnerable groups. Numerous influenza global epidemics, or "pandemics," have been documented, with three occurring in the twentieth century. Some estimates of mortality from the Spanish flu pandemic of 1918-1920 are greater than 50 million people worldwide. Pandemics in 1957 and 1968 together killed more than 1.5 million people (WHO, 1999a).

The World Health Organization (WHO) has developed a sophisticated international program for influenza surveillance and vaccine preparation. Surveillance is maintained by 110 international influenza centers, which continually isolate influenza virus from humans and animals, so that emerging strains are rapidly identified. They provide human isolates to WHO collaborating centers, where the virus is characterized genetically. Results from the influenza network are reviewed biannually, and a recommendation for the antigenic composition of the next year's influenza vaccine is given to vaccine manufacturers (WHO, 1999b). Vaccines for specific strains of influenza are produced and used worldwide, but at present there is too little use of vaccine for it to have an effect on large-scale transmission patterns.

Influenza is transmitted person to person in aerosol droplets, typically through coughing and sneezing. Viruses causing influenza are typed as A, B, or C. Currently there are three different influenza strains circulating worldwide, two subtypes of influenza A and one of influenza B. Influenza type A viruses constantly change, enabling them to evade the immune system of its host, such that people are susceptible to influenza infection throughout life. One mechanism, antigenic "drift," is a series of mutations that occur gradually over time. The other type of change is the more abrupt antigenic "shift," in which a new subtype of the virus suddenly emerges by incorporation of a gene from an animal host influenza strain. Antigenic shift occurs infrequently, but when it happens large numbers of people, and sometimes an entire population, are vulnerable to infection because they have no antibodies that recognize the virus. Influenza B viruses occur almost exclusively in humans, whereas all human influenza A viruses infect avian species, and a few subtypes infect other animals, particularly pigs and horses. There is evidence that the three pandemic influenza strains of the twentieth century arose from the incorporation of genetic material from animal influenza viruses (Webster et al., 1992).

Climate influences. Influenza has a clear seasonal cycle, occurring in North America mainly in late fall, winter, and early spring. The peak number of reported cases averaged over five seasons from 1994 to 1999 occurred in mid-December through February. It is reasonable to assume that the disease transmission cycle is influenced by climate, but the actual driving mechanisms are not well understood and have been the subject of few quantitative studies. Annual influenza outbreaks do not appear to correlate with mean winter or monthly temperature (Langford and Bentham, 1995). The interannual variability in the virulence of influenza strains makes interpretation of the relevant data difficult.

One common explanation for influenza's seasonal cycle is that there is more indoor crowding in the winter, which leads to greater disease transmission. Evidence for the effect of crowding includes the fact that flu epidemics tend to correlate with the start of school and peak during winter holidays, and that outbreaks occur frequently on cruise ships and other "contained" environments. However, laboratory studies of influenza transmission in mice show that, even under identical crowding conditions, flu transmission can still show a seasonal component. This may be due to the effects of humidity on the survival rate of the virus contained in the aerosolized droplets spread by coughing and sneezing (Schulman and Kilbourne, 1963).

Flu is often regarded as a "high-latitude" disease, yet it does occur every year in the tropics. Recent outbreaks occurring on cruise ships have shown that flu can be introduced from the Southern Hemisphere (Australia) and lead to outbreaks even in the middle of the summer (Miller et al., 2000b). More research is needed to gain a better understanding of the basis of the climatic influences on influenza. Particularly useful would be additional animal model studies in which dose and environmental conditions can be controlled.

There are many ways that global warming could conceivably impact influenza transmission. For instance, warming may change bird migration patterns and thus patterns of interaction between humans and infected animals; if warmer weather reduces indoor crowding, this could reduce virus transmission; higher relative humidity and ultraviolet flux could impair virus survival and slow the spread of disease. Other factors that may have greater influence on future transmission patterns include changes in population density, urbanization, and increased air travel.

CRYPTOSPORIDIUM

Disease description. *Cryptosporidium*, one of the most significant causes of waterborne disease in the United States and perhaps throughout the world, is an enteric protozoan that infects the intestinal tract and causes severe diarrhea. *Cryptosporidium* is an obligate parasite; it completes a complex life cycle in the epithelial cells and produces thousands of egg-like structures known as oocysts. The organism is transmitted by the fecal-oral route; individuals become infected when oocysts are washed into water supplies from sewage or animal wastes.

First diagnosed in humans in 1976, it is well recognized as a cause of severe diarrheal illness worldwide (Fayer, 1997). Populations with compromised immune systems are most severely affected, with up to a 50 percent mortality rate reported in some outbreaks (Rose, 1997). Incidence of *Cryptosporidium* infections in the U.S. population varies widely depending on geographic location. In North America there have been 12 waterborne outbreaks of *Cryptosporidium*. The largest outbreak in the United States occurred in Milwaukee, Wisconsin, in 1993, when 400,000 people became ill and 100 died due to contamination of the water supply with fecal wastes (MacKenzie et al., 1994). There is a greater prevalence of infection in populations in Asia, Australia, Africa, and South America; and *cryptosporidium* has also been associated with drinking water outbreaks in the United Kingdom, Japan, and Holland (Smith and Rose, 1998).

Climate influences. The role of climate has not been clearly elucidated in the transmission of *Cryptosporidium*. Data on drinking water outbreaks (from all infectious agents) in the United States from 1971 to 1994 demonstrated a distinct seasonality, a spatial clustering in key watersheds, and a statistical association with extreme precipitation (Rose et al., 2000). This suggests that land use in key watersheds is an important factor, facilitating transport of fecal contaminants from both human sewage and animal wastes into waterways and drinking water supplies during heavy precipitation.

The occurrence of *Cryptosporidium* in surface waters has been reported in 4 to 100 percent of samples examined (Lisle and Rose, 1995). Groundwater, once thought to be a more protected source, has shown between 9.5 and 22 percent of samples to be positive for *Cryptosporidium* (Hancock et al., 1998). Correlations between increased rainfall and increased *Cryptosporidium* oocyst concentrations in river water have been reported (Alterholt et al., 1998). In the Milwaukee outbreak, spring rains and storm runoff were suspected to wash both human and animal wastes into Lake Michigan, overwhelming the drinking water treatment process. The Oxford/Swindon *Cryptosporidium* outbreak in the United Kingdom was also associated with a rainfall event (Rose, 1997), and rainfall was implicated in a waterborne outbreak of giardiasis, a diarrheal disease caused by the similar protozoan *Giardia* (Weniger et al., 1983). Researchers in Brazil suggested that waterborne transmission of *Cryptosporidium* was related to the seasonality of the cases associated with rainfall (Wuhib et al., 1994).

Waterborne disease due to any fecal-oral agent such as *Cryptosporidium* is not only influenced by climate. The incidence of infection in the animal or human population and the type of animal waste handling and sewage treatment will influence the likelihood of oocysts ending up in the environment. The size and hydrology of the watershed and the type and reliability of drinking water treatment will influence the impact on the drinking water. Thus, human, infrastructure, and engineering factors all play important roles in the possibility of waterborne disease.

CHOLERA AND OTHER VIBRIOS

Disease description. Cholera is a diarrheal disease caused by the human pathogenic bacterium, *Vibrio cholera*. Vibrios, including the other pathogenic species—*V. parahaemolyticus, V. vulnificus,* and *V. alginolyticus*—commonly occur in marine and estuarine waters, frequently in association with planktonic copepods. The primary mechanism for transmission of disease is through ingestion of contaminated water or seafood. Chlorination and filtration have effectively eliminated cholera from the water supply in many countries. Currently, the only source of cholera in the United States is contaminated shellfish (fewer than five cases per year).

A cholera pandemic has been ongoing for at least the past 40 years in developing nations of Asia, Africa, and Latin America. In 1998 the number of reported cholera cases worldwide almost doubled. As many as 72 percent of the cholera cases were from Africa (211,748 cases). There was also a dramatic increase in cholera in Central and South America, from 17,760 in 1997 to 57,106 in 1998. Similarly, cases in Asia more than doubled in 1998 compared to 1997, with notable increases in Afghanistan, India, Cambodia, Malaysia, Nepal, and Sri Lanka (WHO, 1999c).

One of the mysteries of cholera epidemics has been that cholera bacteria do not show up in cultures from environmental samples between epidemics; thus, it was difficult to identify the reservoir of bacteria that could initiate a new outbreak. Recently though, immunofluorescent assays have revealed that vibrios are present in the environment even when they cannot be cultured. The vibrios appear to enter a non-culturable phase induced primarily by unfavorable conditions such as low temperature (Colwell and Grimes, 2000). Higher temperatures, in turn, lead to increased numbers of vibrios.

Climate influences. Colwell (1996) hypothesized that a 1990 El Niño event that brought warm waters to the coastal waters off Peru fostered the growth of vibrios and thus contributed to an outbreak of cholera in January 1991 in Peru and neighboring countries (Mata, 1994). Similarly, an association was found between sea surface temperature in the Bay of Bengal and cholera cases reported in Bangladesh (Colwell, 1996). Sea surface height was also found to be associated with cholera outbreaks and may be an indicator of incursion of plankton-laden water inland (Lobitz et al., 2000). More recently, Pascual et al. (2000) carried out a time-series analysis of an 18-year cholera record from Bangladesh and found a significant association between ENSO and the interannual variability of cholera, likely mediated by regional climate variables such as temperature. The confirmation that *V. cholerae* occurs in aquatic environments in association with zooplankton and phytoplankton, and the associations found between cholera cases and sea surface temperature and sea surface height, combine to provide evidence that some cholera epidemics are indeed influenced by climate.

Other vibrios that cause human disease also have growth characteristics dependent on climate. *V. vulnificus*, acquired by eating uncooked or undercooked shellfish, causes primary septicemia and gastroenteritis. *V. vulnificus* proliferates at warmer temperatures and thus could be influenced by climate-induced increases in water temperature. O'Neill et al. (1992) showed that in a New England estuary subject to extreme seasonal temperature fluctuations, there was a strong correlation between levels of *V. vulnificus* recovered from oysters and water temperature and salinity. Motes et al. (1998) investigated the temperature and salinity parameters of waters associated with oysters linked to *V. vulnificus* infections and found that abundance of this pathogen is directly related to water temperature.

Vibrio parahaemolyticus, the second most common vibrio disease in humans after *V. cholerae*, is a frequent food-borne pathogen in Japan. It typically causes acute gastroenteritis when associated with eating uncooked or undercooked shellfish. *V. parahaemolyticus* is the first known example of a human pathogenic bacterium whose growth fluctuates with environmental temperatures. In 1973, Kaneko and Colwell (1973) reported that *V. parahaemolyticus* overwinters in Chesapeake Bay sediment and enters the water column only when temperatures exceed 14°C. They also observed that *V. parahaemolyticus* abundance was proportional to the concentration of zooplankton in the water column, especially copepods. Watkins and Cabelli (1985) observed a similar relationship in Narragansett Bay, Rhode Island. They determined that nutrients associated with waste water stimulated phytoplankton. The phytoplankton supported larger numbers of grazing zooplankton with a resultant increase in the abundance of the associated *V. parahaemolyticus*. From an ecological perspective, it is probable that *V. parahaemolyticus* behaves in a manner similar to that described for *V. cholerae* because both are estuarine bacteria that associate with copepods, both have reasonably strict temperature and salinity requirements, and both can accidentally enter human hosts and cause disease. Hence, it is likely that climate fluctuations conducive to their growth (warmer temperatures and intermediate salinities) lead to increased incidence of disease.

5

Analytical Approaches to Studying Climate/Disease Linkages

A wide variety of approaches are used by researchers to gain an understanding of the relationships between climate and infectious disease. This includes observational-based analyses of *past* or *present* events in nature, model-based projections of possible *future* events, and cross-cutting methods such as risk assessment and integrated assessment. These techniques differ in temporal or spatial extent, source and nature of the data used, and the tools and processes of analysis. The inferences that can be drawn from the different approaches may not be similar in strength or predictive power; some are more relevant to generating hypotheses, others to testing them. Ultimately, each of these different approaches is important in informing the other. The strengths and weaknesses of all of the different approaches are discussed below.

OBSERVATIONAL AND EXPERIMENTAL STUDIES

Observational studies of the relationship between climate variations and health outcomes provide the foundation for developing theory that is used in models and, ultimately, for making forecasts about future impacts associated with climatic changes. As described in the following sections, observational studies can include retrospective and prospective analysis of natural variations, retrospective analysis of historical trends, and interregional comparisons.

Retrospective Analysis of Natural Variations

This approach treats past temporal patterns of climate variability and disease as empirical analogs of future changes. Using time-series analysis of fluctuations or extremes in specific climate variables, historical patterns are compared with those of ecosystem changes or disease outbreaks, with an appropriate time lag sometimes factored in. In recent years this approach has been widely used to study the effects of ENSO phenomenon, especially El Niño events, on infectious disease patterns.

Typically, this analytical approach is employed to look at a single region, so spatial or spatio-temporal variations in the climate/disease relationship are usually not considered. One critical limiting feature of this retrospective approach is that many years of comparable environmental and disease incidence data are usually needed to have confidence that apparent patterns are not occurring by chance. Often, disease surveillance data are inadequate, reducing the applicability of this approach. In addition, the observational nature of this method makes it difficult to separate the influences of other ecological or social changes from those of climate.

For all of these reasons one must be cautious about interpreting the findings of these analog studies and extrapolating the results beyond the specific context of any particular situation. For instance, studies of the effects of an El Niño event do show some of the ways that short-term climate variations can affect epidemic disease, but they are not necessarily a good analog of future long-term climate change. In general, though, this analytical approach does hold potential for improving forecasts of how short-term variability may alter epidemic risk; and if consistent relationships are found over a long time period or in many different places, more confidence can be gained in using these relationships to forecast future changes.

Prospective Observations of Natural Variations

Under some circumstances, surveillance of diseases may be ongoing during periods of anomalous weather events, thus allowing for "prospective" comparison of patterns of variability in disease incidence and climate. Sometimes this involves chance or good fortune in which health surveillance and climate observations at the relevant spatial and temporal scales are already being made to address other questions (e.g., Lindblade et al., 1999). In other cases, intentional focused observations can be planned when a particular climatic event is expected. During the 1998 El Niño, for example, NOAA's Office of Global Programs requested intensified disease and meteorological observations in various sites where surveillance efforts were ongoing, thus creating a prospective sampling and analysis of variability.

A potential pitfall that must be considered in such studies, however, is that one may find higher rates of disease incidence during the period under study, simply as a result of the intensified surveillance efforts. Critical to the interpretation of such observations is a historical record of disease and climate patterns with which to compare each new anomaly. Without such comparison it is impossible to draw inferences from an association of a single event. Prospective observations are needed to test whether hypothesized associations are accurate. These are all reasons why strengthened surveillance of disease incidence is critical to our capacity to analyze future health impacts from climate variability.

Retrospective Analysis of Historical Trends

This approach is similar to that involving retrospective analysis of natural variation, except that it compares the trends or slopes of change during the period of observation. Causal inference is based on the direction and strength of average change over time in both the environmental and disease incidence measures. For instance, Tulu (1996) analyzed malaria trends in the highland region of Ethiopia over the past decade and concluded that increases in nighttime temperatures expanded the altitudinal range of malaria transmission and increased the rate and duration of transmission in areas that were previously epidemic-prone.

This analytical method usually requires long time series of observations, as changes are often slow and interannual fluctuations may confound the trend. Also, the role of climate may be confounded by other trends occurring over the same time period, such as improved sanitation and public health practices; evolution of pathogenicity; land-use changes; or shifts in immunity, age structure, or mobility of the population being studied. However, if similar climate/disease trends are observed in many different regions, this could provide strong insights into underlying relationships and potential future changes.

Interregional Comparisons of Natural Patterns

Comparison of natural spatial patterns in disease incidence and climate is an approach that seeks consistent similarities and differences among regions over a specified time period. If climate patterns are related to disease incidence in a manner that is similar among some regions but different from that observed in other regions, this may help us infer how disease incidence will be affected if the climate in a particular region changes in the direction of that observed elsewhere. An example of this spatial analog approach is a study by Reeves et al. (1994) that took advantage of a 5°C temperature differential between two valleys in California to make inferences about the importance of ambient temperature on the seasonal activity of the main mosquito vector of western equine encephalitis and St. Louis encephalitis.

This construct of inferring possible temporal changes from observed spatial patterns generally assumes that other non-climate conditions are similar in the regions under study. In practice, the role of other environmental and socioeconomic variables is very difficult to identify and disentangle. Different historical patterns of environmental determinants on the disease pattern, and the likelihood of spatial auto-correlation, will complicate reasoning during this process.

Box 5-1
Interdisciplinary Research and Training Needs

The study of climate and human health linkages is highly interdisciplinary in nature, and sustained interdisciplinary research will be absolutely necessary to generate robust understanding of these linkages. While the study of climate itself involves a wide variety of disciplines, these all generally have a common basis in the physical sciences. In contrast, the study of climate/health linkages includes disciplines from the physical, biological, and social sciences. This type of collaboration is generally more difficult because there are fewer common underlying principles and research methods.

There can be numerous impediments to conducting high-quality interdisciplinary research, including difficulties in communicating across disciplines, a dearth of supporting institutional structures such as scientific societies and professional journals, inadequate or nonexistent reward systems at universities and other research institutions, funding structures that are inimical to such research, and biases that value detailed single-disciplinary studies over more generalized interdisciplinary studies.

Interdisciplinary research can be greatly facilitated by the creation and support for university centers and the development of interdisciplinary funding programs. Over the past 30 years or so, many such programs have been developed but in some cases were crippled by a lack of adequate funding. In recent years a few funding programs have been developed specifically to encourage the formation of interdisciplinary research teams, including programs related to the issue of climate and health. These efforts should be encouraged and supported. One concern, however, is that such programs are usually based on three- to five-year funding cycles, which may be too short for developing and executing complex research projects.

Development of long-term interdisciplinary programs offers opportunities for training individuals, including graduate and postdoctoral students, in a variety of methods and activities of value to interdisciplinary research such as learning the basics of disciplines outside one's primary area of expertise, dealing with different types of data analysis, learning how to effectively combine quantitative and qualitative information, and developing flexibility in methodological approaches. This type of training will help ensure that creative interdisciplinary approaches continue to be applied by scientists studying the issue of climate and human health.

Experimental Studies

Our current understanding of the dynamics of many infectious diseases reflects a long history of experimental science ranging in scale from laboratory-based reductionist studies at the molecular and organism levels to field studies at the population level. Manipulative studies are aimed at understanding the mechanism by which particular environmental variables impact parts of an infectious disease transmission system. They can help elucidate parts of a causal pathway and provide information that may be used in combination with other observations or to quantify particular associations.

An example of a field-based experimental manipulation study is an investigation of Rift Valley fever in Kenya carried out by Linthicum et al. (1999). The mosquito vector for Rift Valley fever is known to be associated with flooded areas known as "dambos"; to study the ecology of the disease, researchers intentionally flooded a dambo area and then monitored the resulting insect/virus response. An example of a laboratory-based experimental study is an investigation by Schulman and Kilborne (1963) of the environmental factors that affect influenza transmission, where it was found that the infection rate of mice exposed to the influenza virus was affected by the ambient relative humidity levels.

Interpretation of results from experimental manipulation studies is facilitated when there are multiple observations that can be compared with many "controls." This is usually possible to do in laboratory investigations, but numerous field observations are usually difficult to obtain. As with other field studies, variation in time may be important, making longer studies more likely to produce the most useful information. By controlling for some variables, experimental manipulations limit complex interactions that might actually occur under normal conditions. For these reasons, experimental environmental studies have limited utility in establishing causation or in forecasting. They can be quite valuable, however, in setting parameter values for some processes and in developing hypotheses that are better tested through other approaches.

MATHEMATICAL MODELING

Mathematical modeling can be a powerful tool for gaining insights into the dynamics of infectious disease epidemics, integrating information from laboratory and field studies, providing direction for future experimental and observational studies, evaluating monitoring and control strategies, and making predictions about future disease risk.

As discussed in Chapter 3, the SEIR framework is at the heart of many models of infectious disease transmission dynamics. Yet this is only part of a larger modeling framework that is needed to fully represent the dynamical relationships among climate, ecosystems, and infectious diseases. A comprehensive representation of these relationships needs to account for interactions among factors affecting the disease agent, the vectors and intermediate hosts of these disease agents, the human host (including SEIR parameters), and the environment. Modeling such a complex system presents an extremely challenging task that has yet to be fully met. Progress has been made in recent years, however, in the development of models that simulate parts of this overall system. The different types of modeling approaches used for such studies can be roughly classified into two categories: mechanistic models and empirical-statistical models.

Mechanistic models (or process-based models) use theoretical knowledge of underlying biophysical mechanisms to simulate the health impacts of changes in climate. The models incorporate mathematical equations to represent processes that, in theory, can be applied universally to similar systems in different environments. Mechanistic models are dynamic in that quantitative interactions among multiple variables and feedback processes can be explicitly considered. Forecasted changes in disease risk are based on current interactions of physical and biological variables; thus, most process-based models have not considered various kinds of adaptation or evolution in the many factors that determine transmission or host response. Nevertheless, this approach has been used to forecast how changing climate conditions could lead to ecological changes and affect disease patterns. This approach is useful for exploring alternative scenarios but does not easily convey the uncertainty inherent in forecasts.

Empirical-statistical models are based on relationships between climate and disease-related variables that have been estimated from observational studies. Empirical data on past patterns of variation are used to project how the studied variables may change in the future. This approach may range from applying simple indices of risk (e.g., identifying the minimum temperature threshold for malaria transmission) to using complex multivariate models that consider numerous environmental factors affecting risk. One limitation of this modeling approach is that there may be limited data points to calibrate the projections, making results difficult to validate. Generally, this approach relies little on underlying mechanisms and is not explanatory in nature; often these models are not easily able to consider interactions or other variables not included in the available observations. Regardless, these models can be useful tools for developing hypotheses that indicate where prospective observations will be useful. An advantage of this approach is that empirical-statistical models are often simpler to use and less "data demanding" than mechanistic models.

The two basic classes of models discussed above are not mutually exclusive. Some models use mechanistic approaches to represent the essence of the process under study, combined with statistical approaches to extend the process to the population level. No single approach is clearly superior (or is likely to be sufficient) for creation of reliable predictive models. The most appropriate approaches depend largely on the type of data available and the type of output needed for decision making. In general, however, there are several basic features that are desirable for any type of model used in studies of climate and infectious disease:

• *Small number of parameters.* In general, the simpler a model is, the better we understand its implications. The assumption that more complex models will more closely approximate reality is not justified in cases where there is a paucity of data on relevant "measureables." Models that use a smaller number of parameters by aggregating variables involved in the true process can narrow the range of possible outcomes without implying more knowledge than is available.

• *Well-understood dynamical behavior.* Disease transmission models can be intrinsically nonlinear and may sometimes be chaotic in behavior. Whether

Box 5-2
Modeling the Impacts of Climate Change on Vector-Borne Infectious Diseases

Numerous models have been developed to match the presence of vector species with a discrete range of meteorological parameters and to then project the effects of climate change on vector redistribution (e.g., see Rogers and Randolph, 1993; Sutherst et al., 1995; Lindsay et al., 1998). Discussed below are two recent studies of how future climate change could affect global patterns of potential malaria risk. These studies exemplify the different modeling approaches that can be used to address such questions and also illustrate the widely differing results that can be obtained.

Martens et al. (1999) developed a mechanistic model of the biological processes by which temperature affects mosquito development, feeding frequency and longevity, and the incubation period of the malarial parasite in the mosquito. The suitability of vector habitats is determined by minimum precipitation levels. The model calculates relative changes in the "transmission potential" (the reciprocal of the vector/host ratio necessary to maintain malaria transmission), and the ratio of future to present transmission potential for any given region is taken to indicate the relative degree of future malaria risk. This model predicts that climate change could lead to increases in malaria risk in high-latitude regions (e.g., in Africa and South America), and could also lead to significant increases in malaria risk (from a very low baseline) for much of the United States, Europe, and middle Asia, areas where the malaria vector is currently present but development of the parasite is inhibited by low temperature.

Rogers and Randolph (2000) explored similar questions using a multivariate statistical model. The present-day distribution of malaria was used to empirically establish how the disease is currently constrained by the mean and covariances of meteorological factors including temperature, precipitation, and atmospheric humidity. These statistical relationships were then used to predict potential malaria distribution under future climate change scenarios. Using this approach they estimated only very minor geographical extensions of the potential malaria distribution as compared to the present day; in some areas malaria was predicted to diminish. The reason for these modest changes is that the covariation of the different meteorological factors limits potential expansion of malaria transmission in many regions.

All such models have their specific disadvantages and advantages. The equations in a global model may be inappropriate for particular local conditions. While some models do attempt to take account of local/regional conditions, these studies clearly cannot include all factors that affect species distributions. For example, local geographical barriers and interaction/competition between species are important factors that determine whether species colonize the full extent of suitable habitat. Likewise, the relationship between vector-borne disease incidence and climate variables is complicated by many factors related to a community's socioeconomic status and lifestyle. There is a clear need to validate models on a local or regional scale using historical data. Unfortunately, though, historical data on vector and disease distribution are often not available, especially in many poor countries.

this reflects the actual nature of the system being modeled or is simply an unintended artifact, it is important that the dynamical behavior of a model be well understood.

• *Proper consideration of temporal/spatial scales.* A clear delineation of the temporal and spatial scales at which a model works is essential for its success. Availability of the data at appropriate scales should be taken into account when building a model. For instance, one impediment to using mechanistic models that mimic complex biological and ecological processes is the reliance on micro-environmental parameters that are seldom available in observational data.

• *Incorporation of experimental data and expert opinion.* Scientists have learned a great deal about the biology of various disease vectors through field studies and controlled laboratory experiments. It is important that this information be incorporated into models. Likewise, in cases where there are few experimental data to work with, it is useful to develop modeling approaches that can incorporate qualitative-type information and expert opinions or judgments.

• *Explanatory in spirit but predictive in behavior.* Forecasts generated by disease models will presumably be used for designing effective monitoring strategies and for making public health policy decisions. For such purposes a model that is reasonably good at prediction but somewhat less precise in explanation may still be quite useful. The model has to be appropriate for the purpose but not necessarily a perfect reflection of the complex processes occurring in nature.

There are also numerous issues that must be considered in terms of relating models to available data. Some of the particular challenges that must be commonly addressed in epidemiological modeling are the following:

• *Measurement errors.* Measurement errors are an important concern in many ecological studies. For instance, suppose one wants to develop a population dynamics model to study how mosquito abundance changes over time as a function of environmental variables. It is not possible to actually count all of the mosquitos, but there are various methods that can be used to estimate population abundance. If the model does not explicitly account for the measurement errors associated with these estimates, there can be substantial biases in the inferences drawn from the data.

• *Reporting bias.* This is an important concern in many epidemiological studies. Often, one region will have an effective surveillance/reporting mechanism for a particular disease while other regions do not. It can be difficult to compare the data yielded by these different reporting standards. Sometimes it may be possible to adjust for different reporting standards by modeling the reporting bias explicitly. It is important that reporting biases are either properly estimated or reduced by mandating uniform standards for data collection and reporting.

• *Aggregation bias.* This is another common problem in ecological and public health studies. For instance, in a study of risk factors associated with infant death rates, data analyzed at the city-block level showed a positive correlation between minority population and infant death rate, whereas the same data aggregated at the census-tract level showed a negative correlation (Richard Hoskins, personal communication). This illustrates the importance of considering spatial scale in data analysis and modeling.

• *Data availability and scale.* Although it is often preferable to build models with fine scales of resolution, this requires high-resolution observational data, which is not always available. When information on crucial variables is not available at the appropriate scale, or when data on covariates in a model are available at differing levels of resolution, this can greatly hinder the process of model fitting and validation.

• *Meta-analysis.* One approach that can be used to compensate for a paucity of epidemiological data is to conduct a "meta-analysis" of data pooled from numerous studies. This approach is difficult to apply when there are few studies that cover similar temporal/spatial scales or that focus on similar driving forces and outcomes. However, the field of meta-analysis offers strategies for dealing with some of these difficulties, and in a more qualitative sense, meta-analysis can be useful for looking at differences among studies and providing guidance as to where more testing is needed.

Proper validation is critical for infectious disease models, due to their complexity and potential for nonlinear behavior. But given that all models are only an approximation of a real process, how do we determine whether one model is more appropriate than another? In other words, how does one choose which model is the best approximation of the truth if the truth is unknown? Model validation criteria must balance the fact that a larger number of parameters can mean less reliable parameter estimates on the one hand but more success at mimicking the observed data on the other hand.

Cross-validation procedures provide one means for testing the goodness of prediction. In this approach one of the data points is removed, and the model tries to predict it using the rest of the data. This procedure is repeated for all of the data points, and a summary measure of the performance is reported. The model that has the smallest average squared prediction error is deemed the best in the class. This procedure accounts for the estimation cost associated with the parameters as well as the benefit in prediction due to the complexity of the model.

Developing predictive epidemiological models is an important goal, because "experimental" forecasts of disease outbreaks can provide a valuable means to test our understanding of the linkages among climate, ecosystems, and infectious disease. It is important to recognize, however, that even if a predictive model is feasible from a scientific perspective, that does not necessarily make it appropri-

---

**Box 5-3
Comparison with Other Types of Climate Impact
Modeling Studies**

It is instructive to compare the models used to study climate impacts on infectious diseases with the models used to study climate impacts in other realms. For instance, simulations of climate impacts on agricultural crops and on natural ecosystems typically use dynamic, mechanistic models in which detailed processes are simplified using empirical relationships and aggregation of variables. The fundamentals of these models are generally considered valid over a wide range of contexts, but location-specific input is often required for some parameters. Infectious disease models often use this same general approach, and yet for a variety of reasons, can pose a much more complicated scientific challenge. For instance, infectious disease models must account for numerous social processes that are not constrained by well-defined physical laws and are often highly location specific. Also, in many cases, key relationships between environmental parameters and the parameters characterizing disease agents, vectors, and hosts are not well quantified, either because there are insufficient data to empirically define these relationships, or because the underlying mechanisms linking these parameters are not well understood.

---

ate as a real-world management tool. As discussed in Chapter 7, there can be risks and substantial costs associated with issuing disease warnings and mobilizing intervention strategies. Before decision-makers could use model predictions as a basis for any such actions, the reliability and precision of these predictions would need to be thoroughly understood.

RISK ASSESSMENT FRAMEWORKS

Risk assessment can be defined as the characterization and estimation of potential adverse health effects associated with exposure of an individual or population to hazards. Risk assessment is used by the regulatory community for assessing environmental contaminants and the risks they pose to human and ecosystem health. The National Research Council (NRC, 1983) has defined a four-stage process for risk assessment: hazard identification, dose-response, exposure assessment, and risk characterization. This approach has been used by the U.S. Environmental Protection Agency to develop controls for key waterborne disease agents (U.S. EPA, 1989; Regli et al., 1991; Rose and Gerba, 1991), and has been used to examine the probability of infection and disease resulting from exposure to a variety of pathogenic microorganisms, primarily focusing on fecal-oral and waterborne pathogens (Haas et al., 1999).

Hazard identification refers to both the identification of the microbial agent and the spectrum of human illnesses associated with the specific microorganism (which can range in seriousness from asymptomatic infections to death). Clinical and surveillance data are used to describe what microorganisms are causing what diseases, and quantitative analyses are undertaken to determine the spectra of disease outcomes.

Dose-response refers to the mathematical characterization of the relationship between the dose administered and the probability of infection or disease in the exposed population. In dose-response studies, various doses of specific microorganisms are given to sets of human volunteers, and the percentage of individuals infected at each dose is fit to a best-fit curve. Pathogen/host dose-response data for numerous pathogens have been summarized by Haas et al. (1999).

Dose-response relationships may vary depending on the type of microorganism and the type of exposure. For many pathogens the dose-response relationship is linear only at low doses, and at some point the concentrations are high enough to cause a flattening out of the curve, which reflects very high levels of probability of infection and would result in a large number of symptomatic cases. Most dose-response studies focus on water-borne disease agents, with exposure through an ingestion route. Determining a dose-response relationship for vector-borne diseases may be more difficult. For instance, since high doses of viruses or parasites are likely delivered in a single mosquito bite, the response may simply be dichotomous.

Exposure assessment is aimed at determining the size and nature of the population exposed and the duration of its exposure. This is governed by numerous aspects of a population's overall vulnerability, for instance, how much time is spent in environments where the vector is present, whether homes provide adequate protection from mosquitos and other vectors (e.g., with window screens), and whether there is access to clean drinking water, vaccines, and other public health protections. Exposure assessment also includes characterizing the microorganisms' occurrence (concentration), prevalence (how often the microorganisms are found), and distribution in space and over time, which in turn requires assessing the environmental factors that influence the microorganisms' survival, bioconcentration, and transport.

Quantitative risk characterization is aimed at estimating the magnitude of the public health problem and understanding the variability and uncertainty of the risk. Techniques such as Monte Carlo analysis are used to examine the distribution of exposures and outcomes from the individual to the population level. Dose-response and exposure estimates can be used as input for epidemiological models to describe the transmission in a population and take into account such factors as incubation times, secondary spread, and immunity (Eisenberg et al., 1996).

This framework can potentially be used to study the influence of climate and other environmental factors on disease risk, primarily through their influence on exposure. The first step would be to define and describe the environmental factors that are related directly to the pathogen of concern and the disease outcome. The next step is to evaluate how environmental factors influence the concentration, distribution, prevalence, viability and/or virulence of the pathogen. It is likely that several "probability functions" would be of interest—for instance, the probability that any given dose (e.g., glass of water, mosquito bite) contains the pathogen; the exposure frequency in some unit of time (e.g., number of glasses of water per day, number of mosquito bites per hour); and the distribution of the concentrations of pathogens in the dose(s). In some cases, studies can be designed to relate environmental factors directly to disease or probability of infection, such as the example mentioned earlier where mice were exposed to influenza virus at varying levels of relative humidity to examine how this affected infection rates.

The case of *Vibrio cholerae* provides a useful example of the types of information that might go into an exposure assessment on a climate-sensitive pathogen. *V. cholerae* is spread by the fecal-oral route through contaminated water and food, and as discussed in Chapter 4, there is evidence that *V. cholerae* naturally occurs in estuarine environments in association with copepods. It is hypothesized that the bacteria are then transmitted to humans by copepods getting into water supplies in coastal communities or possibly the food chain. Colwell (1996) has shown that *V. cholerae* can be associated with these zooplankton, and this relationship exhibits a seasonal pattern that can be linked to the seasonal occurrence of cases of disease. Two of the primary environmental variables influencing this complex ecology of pathogenic *Vibrios* are sea surface salinity and temperature. Table 5-1 lists some of the environmental factors that

TABLE 5-1 Environmental and Climate Factors Associated with Exposure to *Vibrio cholerae*

Types of Data/Factors to be Studied	Possible Approaches
Chlorophyll a and/or turbidity associated with rainfall events	Measured by remote sensing
Influence of rainfall and runoff on salinity and nutrients leading to algal blooms. Influence of temperature on phytoplankton growth.	Population dynamic modeling
Influence of phytoplankton on zooplankton dynamics and succession.	Population dynamic modeling
Influence of temperature and salinity on the growth of Vibrio in the copepod	Bacterial growth curves
Concentration of the bacteria in the copepod	Plankton sampling and immunofluorescence microscopy
Numbers of copepods transmitted upstream and in a glass of water	Straining and examination of the water supply

can affect exposure to *vibrios* and the opportunities to monitor and study these factors.

A dose-response model for *V. cholerae* has been developed by Hornick et al., (1971), and this information could be used together with an exposure assessment to estimate the probability of infection and the potential numbers of cases of illness in a population. Such studies, however, must account for the fact that the surveillance and reporting of *Vibrio* cases of illness generally underestimate the true level of infection and risk to the population: and also that time lags of up to several months can exist between environmental changes (such as a rainfall event) and final exposure to the pathogen.

Risk assessment can help quantify the relationships between environmental factors, exposure, and illness. This framework can be used to help determine the combination of factors that contribute to risk and that would be most useful to monitor for the purpose of providing early warnings.

INTEGRATED ASSESSMENT

Integrated Assessment (IA) can be defined as a structured process of using knowledge from various scientific disciplines and/or stakeholders such that integrated insights are made available to decision-makers (Rotmans, 1998). IA has emerged as a new approach for meeting two central challenges of global change research: (1) adequately characterizing the complex interactions and feedback mechanisms among the various facets of global change; and (2) providing support for public decision making through a framework for testing the effectiveness of various policy strategies (Rotmans and van Asselt, 1999). IA provides a coherent framework for working through the causal chain from climate dynamics to climate impacts to policy response strategies. It is an iterative process wherein insights from the scientific and stakeholder communities are communicated to the decision-making community, and, in turn, the evolving informational needs of decision makers provide input for future research.

A wide variety of research methods fall under the rubric of IA. Included are analytical methods traditionally rooted in the natural sciences such as risk analysis and mathematical modeling. Also included are a variety of "participatory" methods rooted in the social sciences, which aim to involve non-scientists as stakeholders in the process and to facilitate stakeholder-scientist interactions.

Chan et al. (1999) developed a conceptual IA framework to identify key linkages among the different systems affected by climate change and that in turn affect epidemiological outcomes; these include changes in transmission biology, ecological changes, and sociological changes. They found that only some of the linkages within this framework have received much attention from the research community. For instance, there have been numerous studies to estimate the direct impacts of temperature on transmission biology, but comparatively little work has been done to integrate climate-related ecological and sociological factors. As noted by the authors, this type of analysis can help identify the different pathways through which infectious diseases are affected by climate and ecological changes, but it does not necessarily help us understand which factors are most critical in any particular context.

There have been numerous IA mathematical models developed for studying global climate change (see Schneider, 1997, or Parson and Fischer-Vanden, 1997, for an overview), including models for assessing climate-related health risks (Martens, 1998). These models try to quantitatively describe as much as possible of the relevant cause-effect relationships (vertical integration) and interactions between different processes (horizontal integration), including feedbacks and societal adaptations. These models integrate global climate change scenarios with local socioeconomic and environmental factors into a coherent modeling framework based on variables describing climate, vectors, parasites, human populations, and health impact. Since infectious disease transmission dynamics are complex systems that can display spontaneous or socially based adaptive responses, some researchers have attempted to develop algorithms that incorporate a capacity for adaptive change and "learning" to simulate such processes (Janssen and Martens, 1997; Sethi and Jain, 1991). It is inevitable that some natural and social phenomena will be oversimplified with such approaches; however, integrated models can provide a useful complement to more focused models that provide highly detailed representation of these complex processes.

As recently summarized in Rotmans and Van Asselt (1999), the primary advantages of IA methods are that they can help to:

• put complex issues in a broader context by exploring their interrelations with other issues;
• assess potential response options to complex problems, including cost-benefit analyses;
• identify and clarify different sources of uncertainty in the cause-effect chain of a complex problem;
• assist in decision making by putting uncertainties into the framework of risk analysis;
• set priorities for future research by identifying and prioritizing knowledge gaps;
• enhance communication between scientists of many disciplines and between scientists and decision makers.

The weaknesses of integrated assessment methods (in particular, of IA models) include the following:

• *High level of integration.* Many processes of relevance to infectious disease dynamics occur at a micro level, far below the spatial and temporal aggregation of current IA models.

• *Inadequate treatment of uncertainties.* IA models are prone to an accumulation of uncertainties that arise from incomplete knowledge about future climatic changes and health-climate relationships, and the inherent unpredictability of future geopolitical, socioeconomic, demographic, and technological evolution.

• *Absence of stochastic behavior.* Most IA models describe processes in a continuous deterministic manner, excluding the potential effects of extreme conditions or "chaotic" events that may significantly influence the long-term behavior of a system.

• *Limited calibration and validation.* The high level of aggregation implies an inherent lack of empirical variables and parameters, and current datasets are often too small and/or unreliable to apply.

SURVEILLANCE/OBSERVATIONAL DATA NEEDS

Epidemiological Data

All of the analytical methods discussed in the previous sections depend highly on the availability of surveillance and observational data. One of the most critical obstacles to improving our understanding of climate/disease linkages is the lack of high-quality epidemiological data on disease incidence for many locations. These data are needed to establish an empirical basis for assessing climate influences and to develop and validate predictive models. Researchers need long-term datasets in order to establish a baseline against which to detect anomalous changes. Likewise, researchers need datasets with broad geographical coverage in order to do comparative studies of how climate affects transmission of disease agents in different contexts.

Currently, most major public health institutions rely primarily on "passive" disease surveillance systems. For instance, U.S. government agencies such as the Centers for Disease Control and Prevention (CDC) and Department of Defense (DoD) rely on state/local public health personnel to submit surveillance data voluntarily (mostly from the records of hospitals and physicians offices); the situation is similar on the international level, where the World Health Organization relies on individual countries to collect and report national data on disease occurrence. In many locations, data are simply not collected since there is no mandate to do so. In other locations, data that are collected are not publicly reported out of a fear that reports of an outbreak will cause a loss of tourism or other economic hardships. This leads to problems of reporting bias, in which disease outbreaks appear worse in some regions simply because more thorough surveillance and reporting are done in these areas. Also, with no universal guidelines or control over how data are collected, it is very difficult to assure quality control or to compare datasets from different sources.

Awareness of these problems and growing concerns about emerging infectious diseases have prompted efforts to develop new programs such as DoD's

Global Emerging Infections Surveillance and Response System. This program and others within the CDC are taking a more active role in monitoring disease incidence by providing standardized criteria for collecting data and developing electronic databases to facilitate rapid standard reporting and sharing of these data. Such efforts should be encouraged on both national and international levels.

Another impediment is that disease surveillance data collected for particular scientific studies are usually considered to be the property of the researchers who collected them and only become available if and when published in the scientific literature. There are a variety of steps that can be taken (and in some cases are already in place) to encourage investigators to make their datasets publicly available. For example, funding agencies can require that research proposals include a plan for data dissemination, scientific journals can make data accessibility a condition for publication of a study, and submission of data to a central database could be counted as an official publication for the researcher.

Technological developments such as the advent of the Internet and steadily decreasing costs of computing and communications are providing exciting new opportunities for sharing data among researchers all over the world. Yet at the same time there are political and economic developments that jeopardize the open exchange of data among scientists, especially across national boundaries. Data exchange is affected by governmental concerns related to national security, foreign policy, and international trade and an increasing push toward commercialization of electronic databases. These constraints may present serious difficulties in attempts to study regional- and global-scale infectious disease patterns.

A more daunting challenge is the fact that many countries lack the resources to maintain even the most basic population-based disease surveillance programs. For the foreseeable future, the establishment of comprehensive epidemiological surveillance is not likely to be an attainable goal in such regions. It may be possible, though, to target resources for intensive surveillance in specific locations where the populations are most vulnerable to infectious disease threats. Such programs may require substantial assistance from international organizations, but in order to maintain surveillance systems over the long term, it is important to develop each country's internal capacity and intellectual/technological infrastructure.

It must also be recognized that for many diseases even the best surveillance systems will capture only a fraction of the infections that occur in a given population. For instance, in developing countries most individuals who develop dengue fever never seek care and those who do are rarely tested to confirm or rule out dengue. Likewise, individuals with cholera or malaria in many developing countries often do not seek medical care. For the purpose of understanding climate/disease linkages, however, much can still be learned by studying *relative* changes and patterns of disease transmission. Surveillance programs that capture only a fraction of the actual disease incidence can still provide useful information as long as consistent data collection methodologies are used over time or across regions.

Environmental Observational Data

In order to relate changes in disease incidence to environmental changes, epidemiological surveillance must be coordinated with meteorological and ecological observations. Ground-based measurements of meteorological parameters such as daily mean temperature, daily accumulated precipitation, and relative humidity are collected at thousands of sites around the world. There are still some regions of the world, however, where weather observations are quite limited or where there have been significant declines in meteorological reporting in recent years. It is important that these networks be maintained, as they provide vital information both about climate/disease linkages and about climate variability itself.

For the past several decades, meteorological observations have been processed and archived on a global basis through a system of world/regional meteorological data centers. The World Weather Watch, part of the United Nations' World Meteorological Organization, is responsible for planning and coordination of the system, but the national meteorological system of each member nation is responsible for actually operating the data collection and processing centers. This type of international system for collecting and sharing scientific data may provide a good working model for the public health community to follow.

Routine observations of ecological parameters such as soil moisture, vegetative cover, and sea surface temperature are rare, but new remote-sensing technologies are rapidly expanding the opportunities to monitor many of these parameters on a global scale (as discussed further in the following section). However, disease dynamics are predominantly driven by environmental factors at fine spatial and temporal scales (e.g., mosquitos are affected more by temperature in their immediate microenvironment than by mean ambient conditions), and thus methods must be developed for relating remote-sensing data to conditions occurring on these smaller scales.

For both environmental monitoring and public health surveillance, a central challenge is maintaining observations of key parameters over the course of years to decades. Such extended time series are needed in order to establish a baseline against which one can detect long-term trends or patterns of variability.

Remote-Sensing Surveillance Tools

In recent years investigators have begun using remotely sensed data collected by satellite-borne instruments to study environmental factors that can influence disease transmission risk for particular regions. Over the next several years dozens of new instruments will be launched that will provide high-resolution monitoring for a wide variety of environmental parameters relevant to disease risk (Beck et al., 2000).

One of the instruments being used in this capacity is the Advanced Very High Resolution Radiometer (AVHRR), which is carried aboard NOAA-series satellites. AVHRR can be used to calculate a Normalized Difference Vegetation Index (NDVI, or "greenness index"), a tool for monitoring changes in vegetation growth (Justice et al., 1985). This is a widely useful parameter to monitor, since nearly all vector-borne diseases are linked to the vegetated environment during some aspect of their transmission cycle. AVHRR has been in continuous operation since 1981, and this multi-decade time series makes it possible to calculate statistically significant variations from the long-term mean value. AVHRR provides data with high enough resolution (1 kilometer) to be useful for studies on fine spatial scales. Scientists have used NDVI data to identify areas of potential Rift Valley fever activity in regions of eastern Africa (Linthicum et al., 1999), and similar techniques have been used to study other diseases such as trypanosomiasis, schistosomiasis, and malaria.

Soil moisture is an important parameter for evaluating the suitability of a particular habitat for disease vectors such as mosquito larvae, ticks, and snails. NDVI provides an indirect proxy for soil moisture, but instruments that can directly measure soil moisture (e.g., microwave sensors, synthetic aperture radar) also are in operation. Instruments that can "see through" heavy vegetation will be especially useful since, at present, soil moisture for regions such as tropical rain forests can be obtained only by surface sensors.

Remote-sensing data can also be used to aid the study of water-borne disease. For instance, Lobitz et al. (2000) have studied the relationship between cholera outbreaks and sea surface temperature (from the NOAA/AVHRR instrument), sea surface height (from the TOPEX /Poseidon radar altimeter), and chlorophyll concentration (from the OrbView-2/SeaWIFS instrument). Satellite instruments such as SeaWifs are also used to provide information about harmful algal blooms and toxic elements that can infect shellfish.

In recent years several major programs have been established to help foster the use of remote sensing technologies by the health care community, such as MALSAT (Environment Information Systems for Malaria and Meningitis) of the Liverpool School of Tropical Medicine, the Environment and Health Initiative of NASA's Earth Science Enterprise, the INTREPID (International Research Partnership for Infectious Diseases) program, and NASA's Center for Health Applications of Aerospace Related Technologies (CHAART). Through a series of workshops involving scientists from the environmental science and health care communities, CHAART has compiled a comprehensive list of remotely sensed factors that can help in the study of disease (see Table 5-2) and the current/planned instruments that will measure these parameters.

Geographic Information Systems (GIS), a framework for analyzing geographically referenced data, can greatly facilitate integration of remotely sensed parameters with health data. For instance, maps of potential vector habitat obtained by analysis of remote-sensing data can be "layered" with geographically referenced data on land use, human population, and so forth, to create maps of disease risk. For further discussion of the ways in which these technologies can be applied in the field of epidemiology, see Hay et al. (2000).

TABLE 5-2 Potential Links Between Remotely Sensed Factors and Disease

Factor	Disease	Mapping Opportunity
Vegetation/crop type	Chagas disease	Palm forest, dry and degraded woodland habitat for triatomines
	Hantavirus	Preferred food sources for host/reservoirs
	Leishmaniasis	Thick forests as vector/reservoir habitat in Americas
	Lyme disease	Preferred food sources and habitat for host/reservoirs
	Malaria	Breeding/resting/feeding habitats; crop pesticides vector resistance
	Plague	Prairie dog and other reservoir habitat
	Schistosomiasis	Agricultural association with snails, use of human fertilizer
	Trypanosomiasis	Glossina habitat (forests, around villages, depending on species)
	Yellow fever	Reservoir (monkey) habitat
Vegetation green-up	Hantavirus	Timing of food sources for rodent reservoirs
	Lyme disease	Habitat formation and movement of reservoirs, hosts, vectors
	Malaria	Timing of habitat creation
	Plague	Locating prairie dog towns
	Rift Valley fever	Rainfall
	Trypanosomiasis	Glossina survival
Ecotones	Leishmaniasis	Habitats in and around cities that support reservoir (e.g., foxes)
	Lyme disease	Habitat for deer, other hosts/reservoirs; human/vector contact risk
Deforestation	Chagas disease	New settlements in endemic-disease areas
	Malaria	Habitat creation (for vectors requiring sunlit pools)
		Habitat destruction (for vectors requiring shaded pools)
	Yellow fever	Migration of infected human workers into forests where vectors exist
		Migration of disease reservoirs (monkeys) in search of new habitat
Forest patches	Lyme disease	Habitat requirements of deer and other hosts, reservoirs
	Yellow fever	Reservoir (monkey) habitat, migration routes
Flooded forests	Malaria	Mosquito habitat
Flooding	Malaria	Mosquito habitat
	Rift Valley fever	Flooding of dambos, breeding habitat for mosquito vector

	Schistosomiasis	Habitat creation for snails
	St. Louis encephalitis	Habitat creation for mosquitos
Permanent water	Filariasis	Breeding habitat for Mansonia mosquitoes
	Malaria	Breeding habitat for mosquitos
	Onchocerciasis	Simulium larval habitat
	Schistosomiasis	Snail habitat
Wetlands	Cholera	Vibrio cholerae associated with inland water
	Encephalitis	Mosquito habitat
	Malaria	Mosquito habitat
	Schistosomiasis	Snail habitat
Soil moisture	Helminthiases	Worm habitat
	Lyme disease	Tick habitat
	Malaria	Vector breeding habitat
	Schistosomiasis	Snail habitat
Canals	Malaria	Dry season mosquito-breeding habitat; ponding; leaking water
	Onchocerciasis	Simulium larval habitat
	Schistosomiasis	Snail habitat
Human settlements	Diseases	Source of infected humans; populations at risk for transmission
Urban features	Chagas disease	Dwellings that provide habitat for triatomines
	Dengue fever	Urban mosquito habitats
	Filariasis	Urban mosquito habitats
	Leishmaniasis	Housing quality
Ocean color	Cholera	Phytoplankton blooms; nutrients, sediments
Sea surface temperature	Cholera	Plankton blooms (cold water upwelling in marine environment)
Sea surface height	Cholera	Inland movement of Vibrio-contaminated tidal water

SOURCE: adapted from Beck et al., 2000

6

Temporal and Spatial Scaling: An Ecological Perspective

Over the past several years scientists have engaged in a wide array of investigations aimed at understanding the ecological consequences of climatic changes occurring over different temporal and spatial scales. Through these studies a great deal has been learned about the confounding methodological issues that arise when data characterizing climate impacts at one temporal or spatial scale are used to draw conclusions about potential impacts on a different scale. Such lessons are highly relevant to efforts to predict how disease pathogens and vectors will respond to climatic changes. In particular, insights from ecological studies can help us identify both the opportunities and the potential pitfalls of using studies of climate variability to predict disease impacts of long-term anthropogenic climate change.

BIOLOGICAL EFFECTS OF OBSERVED CLIMATE VARIABILITY

As discussed in Chapter 3, climate varies naturally on a wide range of temporal and spatial scales, and over the past century, the global climate has been gradually warming. Climate can also be manipulated under controlled experimental conditions to achieve variability at prescribed time scales. The following paragraphs provide an overview of the range of ecological phenomena that are observed to vary in response to these different forms of observed climate variability.

There are numerous ecological changes associated with *spatial climate gradients*. For instance, associated with latitudinal and altitudinal climate gradients are dramatic changes in soil fertility, species composition, growth rates, and timing of reproductive cycles. The ecological consequences of regular *diurnal and seasonal climate variability* are also well studied and relatively predictable. Day/night and summer/winter temperature differences strongly constrain ecological processes and the population dynamics of disease vectors. Both excessively cold nights/winters and excessively warm days/summers can limit the geographic ranges and reproductive rates of insects as well as affect the rates of microbial activity in soil and water. Examples of plant and animal adaptations to diurnal and seasonal climate variability abound, including migration, hibernation, nocturnality, reproductive cycles, torpor, and leaf shedding. The constraining effects of diurnal and seasonal climate variability on the geographic ranges and the growth and reproductive rates of organisms are intertwined because the hot extreme comes in summer daytime and the cold extreme in winter nighttime.

Interannual to decadal scale climate variability results from coupled atmosphere-ocean processes, and possibly from sunspot cycles and other as yet unexplained driving forces. The ecological and health consequences of these variations can be considerable. For example, climatic variations associated with the ENSO, with multi-year droughts, and with multidecadal monsoon and hurricane cycles are known to correlate with vegetation productivity, bird-nesting success, abundance of insects, and numerous other ecological parameters (e.g., Levins et al., 1994; Tucker et al., 1991). On much longer time scales, the ecological changes caused by the *"little ice age"* and *glacial-interglacial transitions* and longer-term climate changes can be enormous, including huge range shifts or even extinction of species (Campbell and McAndrews, 1993; Davis and Zabinski, 1992; Webb, 1986).

A wide variety of ecological trends are statistically associated with the *long-term warming trend* that has been occurring over the past century. These include earlier arrival of spring (as marked by biological events such as egg laying and vegetation flowering); bird, mammal, and amphibian population declines; and range shifts in butterflies, birds, and marine invertebrates (Brown et al., 1999; Crick and Sparks, 1999; Grabherr et al., 1994; Myeni et al., 1997; Parmesan, 1996; Thomas and Lennon, 1999). Evidence that these changes are actually caused by trends in climate is compelling in some cases (e.g., observed trends in plant-flowering phenology) and at least suggestive in most of the other cases. At many locations, nighttime and winter air temperatures during the past 100 years have increased more than have daytime and summer temperatures. These reductions in the amplitudes of diurnal and seasonal cycles represent potential influences on ecological phenomena, independent of any effects of a change in mean temperature. Organisms are generally more sensitive to temperature extremes than they are to mean temperature, and these amplitude changes could generate geographic-range shifts and altered population densities of a variety of species. The duration of extreme temperature episodes can in some cases be as ecologically important as the extreme temperatures reached during such episodes.

Climate can also be altered by deliberate *experimental manipulation* to learn about ecological responses to these alterations. Climate manipulation experiments can endure for periods from less than a year to over a decade. They can be carried out either in small laboratory chambers or in field plots of varying sizes. In these experiments, soil and/or air temperature are manipulated with the use of overhead heat lamps or heating wires in the soil or with passive devices that serve as small open-top greenhouses (Shen and Harte, 2000). In some field experiments precipitation is also manipulated, and in a few cases temperature increase is combined with an increase in ambient carbon dioxide concentration to more accurately simulate the conditions of future atmospheric conditions. Experiments of these types can affect the relative growth rates of plant species, the population of invertebrates, the timing of plant reproductive cycles, biogeochemical cycling rates, and the quantity of carbon stored in organic form in soil (Harte and Shaw, 1995; Saleska et al., 1999).

CONFOUNDING INFLUENCES ON ECOLOGICAL FORECASTING

By investigating how a target parameter such as disease vector abundance responds to observed climate variability, it may be possible to make reliable deductions about how that target parameter will respond to future anthropogenic global climate change (AGCC), either by direct statistical extrapolation of the observed relationship or by a more complex analysis based on a mechanistic understanding of these relationships. Here we discuss the factors influencing the validity of the conclusions drawn from such investigations and the opportunities that this approach to prediction provides. Our purpose is to provide a framework for evaluating the suitability of the different empirical and theoretical approaches that have been proposed to predict the impacts of AGCC.

• *Mismatches in temporal scale impede efforts to predict.* Ecological adaptations to glacial-interglacial and other long-term climate cycles occur on a time scale that is much longer than the time scales of concern for AGCC. Since climate is not the only environmental parameter that changes over these long time scales, ecological responses to AGCC are not necessarily predictable based on the insights gained from paleoclimatic investigations. For instance, factors such as soil quality constrain plant species composition and growth rates, and the mechanisms shaping soil quality operate over time scales that are relatively long compared to the time frame of AGCC. Thus, over sufficiently long time intervals, plant distributions are likely to correlate with climate change, but those correlations may not provide much insight into short-term responses of plants to AGCC.

Attempts to deduce the potential impacts of AGCC based on the impacts of diurnal, seasonal, interannual, or decadal climate variability suffer from related problems, since in these cases the time frame is inappropriately short compared to that characterizing AGCC. Many organisms' traits are genetically adapted to short-term diurnal and seasonal climate cycles and are not likely to provide reliable indication of responses to AGCC. For instance, ecological responses to the ENSO cycle may not always be indicative of responses on the time scale of AGCC. Changes in ecosystem productivity that are observed to accompany the ENSO cycle are, in part, constrained by the species composition of these ecosystems; and major transitions in species composition occur over century to millennial time scales, not in response to individual El Niño/La Niña events. Finally, extrapolation across these different time scales can be confounded by the fact that anthropogenic stresses on ecosystems such as land-use changes, resource exploitation, and population growth are likely to change more significantly over the course of decades to centuries than over seasonal to interannual time scales.

• *Correlation in time does not imply causation.* The problem of establishing causation besets efforts to deduce future responses to AGCC from observations of ecological responses to the past hundred years of climate change. The correlations between ecological and climatic time series could be coincidental if, for example, third-party factors such as land-use changes are simultaneously driving both climate and ecosystem change. Only manipulated climate change experiments provide an unambiguous way to distinguish causality from correlation; but such experiments are intrinsically limited to temporal and spatial scales that may be too small to provide reliable predictions.

• *Correlation in space does not imply causality at AGCC time scales.* By examining the spatial correlation between climate variations and ecosystem parameters, deductions can be made about how ecosystems will respond to changes in climate over time. The validity of those deductions, however, depends on two premises. The first is that the changes in ecological parameters along spatial gradients are actually driven by the associated climate gradients. The second, called the "space-for-time" assumption, is that the mechanisms of ecological change over time (on decadal to century time scales characteristic of AGCC) are sufficiently similar to the mechanisms that create ecological gradients resulting from spatial climate variability. Because the latter operate over much longer time scales (typically many centuries to millennia) than the former, and since climate is not the only environmental parameter that varies along spatial climate gradients, the space-for-time assumption is not necessarily valid, although it does appear to hold true in some cases. For example, experimental studies have shown that variations in plant phenology (timing of the reproductive cycle) along an elevational gradient in montane meadow habitat provide a remarkably accurate prediction of the phenological response of plants to manipulated climate warming (Price and Waser, 1998). On the other hand, the responses of a number of other ecological variables (including soil organic matter, plant productivity, and species composition) to manipulated climate change were not well predicted by their patterns of change along spatial climate gradients (Dunne, 2000).

•   *Ecological and climatic phenomena are dependent on spatial scale.* Because of the relatively small size of experimental field plots and laboratory chambers used for climate manipulation experiments, observed ecological responses may be quite different from those that would occur in large ecosystems. For example, studies of the influence of climate change on soil fertility and plant productivity on experimentally warmed plots of area about 10 m^2 could not possibly capture the influence on plants and soil of changing populations of large herbivores.

•   *Linear extrapolation is often misleading in ecology.*   Ecological phenomena often depend on climate parameters in highly nonlinear ways, and thus the differences in magnitude between natural climate variability cycles and AGCC can confound efforts at prediction. The mismatch in time scales necessitates the application of reliable interpolation procedures to predict the effects of a relatively small change in temperature (AGCC) based on understanding of the effects of a larger change (e.g., a diurnal or seasonal cycle). For example, many insects in temperate or cold climates are active during daytime and go into nighttime cold-induced torpor. The dependence of insect nighttime activity rates on air temperature is not well characterized, but is thought to be a threshold relationship. Thus, it is unlikely that a simple linear interpolation of activity between day and night temperatures would yield reliable predictions of activity rates at the new nighttime temperatures resulting from AGCC (which would likely be intermediate between current day and night temperatures).

For all of the reasons discussed above, ecological studies point to both pitfalls and possibilities in the use of observed climate variability data to forecast the effects of AGCC. Each type of dataset possesses shortcomings that reduce the opportunity to draw unambiguous conclusions. Were a combination of such datasets (e.g., from interannual variability, spatial gradients, and responses to experimentally manipulated climate) all to point to a single consistent conclusion about ecosystem response to climate variations, this would greatly enhance confidence in the predictions about responses to AGCC. Even when results from different kinds of measurement yield diverging predictions (because of differing spatial and temporal scales), the use of models that explicitly include scale-dependent mechanisms may allow reconciliation of the differing conclusions and provide the insight needed to draw defensible predictions.

The types of scaling difficulties faced in the study of ecological changes are highly relevant to the study of infectious diseases, especially in cases where the transmission cycle is closely associated with ecological changes. For instance, in the urban dengue system breeding occurs primarily in man-made containers, an "environment" that isn't expected to change significantly in response to climatic changes. Under these circumstances, responses to ENSO events or to artificially manipulated climate could provide reliable insight into the potential consequences of AGCC. In contrast, in the trypanosomiasis/tsetse fly system of sub-Saharan Africa, vector dynamics are critically influenced by natural ecosystems that are subject to uncertain change in the face of AGCC, and in this example there may be scaling problems inherent in using observed climate variability responses to anticipate AGCC responses. Malaria may present an "intermediate" example, since the environments that foster its transmission range from quasi-stable ecosystems (e.g., rice-growing regions) to more complex situations where vector dynamics are intimately tied to the natural environment.

7

Toward the Development of Disease Early Warning Systems

An early warning system is an instrument for communicating information about impending risks to vulnerable people before a hazard event occurs, thereby enabling actions to be taken to mitigate potential harm, and sometimes, providing an opportunity to prevent the hazardous event from occurring. Early warning systems are routinely used for hazardous natural events such as hurricanes and volcano eruptions. In contrast, to date very little attention has been paid to the development of such systems for infectious disease epidemics. The goal of a disease early warning system would be to provide public health officials and the general public with as much advance notice as possible about the likelihood of a disease outbreak in a particular location, thus widening the range of feasible response options. The inherent dilemma of an early warning system, however, is that more lead time usually means less predictive certainty, as illustrated in Figure 7-1.

The most commonly used bellweather of an impending epidemic is the appearance of early cases of the disease in a population. In some instances, "sentinel" animals are placed in high-risk locations and monitored for evidence of infection, since infections among these animals will typically presage human cases. These "surveillance and response" approaches provide fairly high predictive certainty of an impending disease outbreak but often leave public health authorities with little advance notice for mobilizing actions to prevent further spread of the disease agent.

In contrast, ecological observations and climate forecasts can potentially be used in efforts to predict the appearance of a pathogen and thus allow opportunities to minimize its transmission. This approach is likely to have a much lower

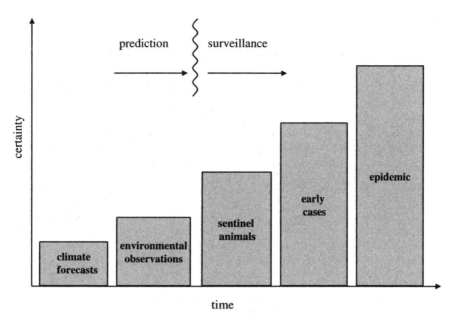

FIGURE 7-1 Progression of the types of information that can be used to indicate an impending disease outbreak.

predictive value, however, given the uncertainties associated with most climate/disease relationships and the confounding influences of other factors. It is highly unlikely that precise predictions of an epidemic could be made solely on the basis of climate forecasts and environmental observations. Yet, this information can feasibly be used as the basis for issuing an alert (or a "watch") that environmental conditions are conducive to disease outbreak, which in turn can trigger intensive surveillance efforts for the area in question. If surveillance data then confirm the presence of the pathogen or an increase in its abundance subsequent warnings could be issued as needed. This watch/warning approach is analogous to the system used by the U.S. National Weather Service to alert communities to impending severe weather. A benefit of this multi-staged early warning approach is that response plans can be gradually ramped up as forecast certainty increases. This would give public health officials several opportunities to weigh the costs of response actions against the risk posed to the public.

DEVELOPING EFFECTIVE EARLY WARNING SYSTEMS

Early warning systems are often defined narrowly as instruments for detecting and forecasting an impending hazard event, but this definition does not clar-

Box 7-1
Approaches to Assessing Disease Risk

A variety of tools/methods can be used to assess the risks posed to a community by climate and ecological changes and to evaluate strategies for responding to those risks. The following are examples that illustrate that the most useful approach depends highly on the context of each particular situation.

Sophisticated mathematical models can be particularly useful in situations where there is a wide array of environmental and other variables relevant to transmission of a disease agent and where it is desirable to weigh the benefits of different intervention strategies. For instance, a model of Lyme disease transmission developed by Mount et al. (1997) simulates the effects of climate and numerous other factors (such as habitat type, host type, and density of tick populations) on disease risk, and this model is used by public health officials to study the potential effects of different strategies for controlling disease vector populations.

In cases where there is known to be a relatively straightforward and consistent association between environmental conditions and disease outbreak, the most useful basis for assessing risk may be a simple extrapolation of past statistical associations. For instance, it is known that Rift Valley fever outbreaks are consistently associated with heavy rainfalls in some locations, and that meningococcal meningitis outbreaks are associated with dry, dusty weather in other locations. This information can provide a simple qualitative guide that "front-line" public health officials may sometimes find more useful than complex model analyses.

Combinations of model-based and simple empirical approaches can also be effective in some contexts. For instance, a dengue early warning system is under development in which a mechanistic SEIR model is used to estimate dengue transmission thresholds for a particular community in terms of ambient temperature and immunity status of the community (Focks et al., 2000). Field surveys are then conducted to estimate the number of mosquito pupae per person in the community and the proportion of the transmission threshold produced by different types of water filled containers (i.e., mosquito-breeding sites). Together, this information allows the creation of simple "look-up" tables for normal, El Niño, and La Niña years, which are used to estimate a community's vulnerability to epidemic dengue transmission and to determine optimal strategies for eliminating breeding containers.

ify whether the warning information is actually used to reduce risks. To fulfill an effective risk reduction function, an early warning system should be understood as an information system designed to facilitate decision making of the relevant national and local-level institutions and to enable vulnerable individuals and social groups to take actions to mitigate the impacts of an impending hazard. The focus must be not only on improving hazard monitoring and prediction but also on improving coordination among relevant parties, such as the scientific organizations that forecast hazard events, the national and local management agencies that assess risk and develop response strategies, and the public communication channels used to disseminate warning information.

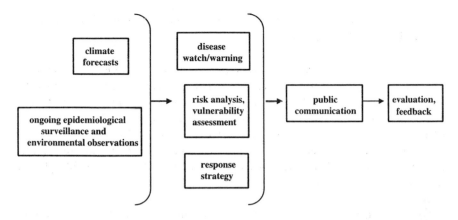

FIGURE 7-2   Diagram of the components of an effective disease early warning system.

Figure 7-2 illustrates the different components necessary for an operational early warning system.   *Climate forecasts* and information from ongoing *epidemiological surveillance* and *environmental observations* may be used as input for predictive models that generate *watches* or *warnings* about an impending disease risk.   This information is then coupled with *vulnerability assessments* to determine which segments of a population are most likely to face harm from an impending hazard and *risk analysis* to determine the likely impact of the impending hazard on these groups.   The actions required to reduce the impacts of an impending hazard are determined through development of a *response strategy*.   Finally, a *public communication system* facilitates the timely dissemination of information on impending hazards, risk scenarios, and preparedness strategies to vulnerable groups.   Issues related to climate forecasting and predictive disease models are discussed in earlier chapters.   The other central components of an early warning system are discussed in more detail below.

Epidemiological surveillance. Epidemiological surveillance systems that are ongoing and systematic, that use standardized routines for quality assurance, and that provide for analysis and timely dissemination of information are critical to early warning systems.   The characteristics and limitations of current surveillance systems are discussed in Chapter 5.   Historically, surveillance has focused specifically on monitoring the incidence of infection or disease.   In the context of an early warning system, however, surveillance also needs to include monitoring of changes in vector population abundance, and may entail the use of "sentinel" animal networks (such as chicken flocks) to provide an early indication of the pathogen's presence in a particular area.   These types of ecologically-based surveillance systems are already commonly used in some contexts, for example, in surveillance for St. Louis encephalitis and Western Equine encephalomyelitis in California (Reisen et al., 1992; Reisen et al., 2000) and West Nile virus in the eastern U.S. (MMWR, 2000).

Environmental observations. Systematic climate observations are an important component of an early warning system, in part because the impacts of weather or climate events often depends on antecedent conditions. For instance, whether a heavy rainfall event will lead to flooding may depend on the recent history of precipitation in the area. New remote-sensing technologies for monitoring ecological parameters such as soil moisture, vegetative cover, and sea surface temperature on a global scale (discussed in Chapter 5) will inevitably play an important role in disease early warning systems.

Vulnerability assessment. Vulnerability refers to a population's sensitivity to a hazard, as well as its ability to cope with the hazard. Vulnerability assessment provides a context for interpreting surveillance data and for understanding the impacts of disruption among any of the links that connect nutrition, shelter, economic systems, and human health. At the household level, factors such as diet, shelter, sanitation, and water supply affect vulnerability to infectious disease. Common risk factors in developing countries (and in parts of developed countries) include lack of access to clean water, poor sanitation, inadequate shelter, and low immunization coverage. A community's vulnerability is also affected by such demographic factors as age structure and population density, and by pre-existing health threats such as HIV. Vulnerability assessment should be closely coupled to public health surveillance; both are crucial for evaluating the potential for outbreaks and for developing disease control strategies.

Risk analysis. Risk analysis is carried out to assign specific probabilities to the likely impacts of an impending hazard. Risk is a complex variable related to hazard types and patterns of vulnerability, possible impacts of the hazard events, and the capacity of communities to absorb and recover from these impacts. In many countries, risk patterns are rapidly changing as a result of urbanization, economic change, population growth, migration, environmental degradation, and armed conflict. Yet risk analysis is still fundamentally static in character in many places, often because cartographic and census information is so out of date that it bears little relationship to real risk levels. A priority must be to build the capacity to monitor dynamic changes in risk patterns at a high level of spatial and temporal resolution, in order to provide accurate information on the likely impact of a specific hazard event in a given area.

Preparedness/response. Warning systems must be developed in concert with developments in local-, national-, or regional-level response capabilities, particularly in highly vulnerable areas. Often, improvements in scientific predictive capabilities are not accompanied by commensurate improvements in the ability to actually use this warning information. Experience suggests the importance of developing response strategies based on community needs and priorities. Response plans that contradict the accepted coping strategies of a vulnerable group are not likely to be implemented effectively. The actions taken to prevent or mitigate disease outbreaks can often carry significant costs and, in particular, can pose a major burden on resources in poor countries. If the costs of carrying out a recommended response strategy are seen to be greater than the public health benefits gained, this can undermine the credibility of the early warning system as a whole. In addition, implementation of control measures (such as widespread spraying of insecticides) can sometimes face active public resistance. Such potential pitfalls and cost-benefit considerations need to be carefully assessed so that response plans can be adapted to best suit the priorities, needs, and capacity of the local community.

Public communication. To ensure that warning information and recommended response strategies are heeded by the populations at risk, effective public communication strategies must be developed. As discussed in Freimuth et al. (2000), effective health communication programs identify and prioritize audience segments; deliver accurate, science-based messages from credible sources; and reach audiences through familiar channels. Attention must be paid to issues

---

Box 7-2
Examples of Response Strategies

Specific strategies for responding to a disease warning depend on the perceived magnitude of the threat and on the socioeconomic conditions and institutional resources of the region in question. As an example, the following is a list of actions that were recommended or supported by the Pan-American Health Organization to help countries in South and Central America prepare for the onset of the 1997-1998 El Niño (adapted from PAHO, 1998):

• Training workshops for public health professionals in high-risk areas for strengthening entomological surveillance, vector control, and prevention activities such as the distribution of mosquito netting.
• Provision of basic supplies for water storage and treatment.
• Local workshops to find solutions to environmental sanitation problems.
• Identification of shelter sites and requirements for their installation and control of food distribution.
• Characterization of rodents and vectors of public health significance in disaster areas.
• Strengthening of laboratory diagnosis for leptospirosis and hantavirus.
• Vaccination of the affected population against whooping cough, tetanus, and diphtheria to guard against potential outbreaks.

of credibility and trust. If the warning authority lacks status in the area concerned or has a history of conflicting relationships with local communities, warnings may not be well received. Likewise, if too many warnings had been given previously or no warning was given at all when previous disease outbreaks did occur, the credibility of future warnings will be jeopardized. Finally, to prevent overreaction and even panic by the general public, disease early warnings must include clear explanations of the actual level of risk involved and of the particular groups that are most vulnerable.

Implementation Issues

Early warning systems are only as good as their weakest link, and they can fail for a number of reasons: the hazard forecast or risk scenarios generated may be inaccurate; there may be a failure to communicate warning information in sufficient time or in a way that can be usefully interpreted by people exposed to a hazard; appropriate response decisions may not be made due to a lack of information, opaque decision-making procedures, or perceived political risks; or intervention actions may incur unacceptable costs for the community.

One potentially key barrier in effectively implementing an early warning system is lack of clear decision-making protocols among the relevant institutions. In any community there needs to be a designated focal point for coordinating the actions of key institutions such as water treatment plants, hospitals, public health agencies, and transportation providers. In this respect, public health organizations could build on lessons that have been learned by organizations that coordinate responses to natural disasters, such as the U.S. Federal Emergency Management Agency. Clear performance standards and regular performance evaluations can help build public awareness and confidence in the system. To help ensure that lessons from one event are incorporated into planning for future events, management agencies need to evaluate the extent to which warnings are heeded and preparedness strategies are implemented.

A critical lesson that has been learned from existing early warning systems is the importance of involving the system's end users (public health officials, mosquito control officers, water quality monitors, policymakers, etc.). "Stakeholder" participation will enable those who generate forecasts to understand what is needed in terms of lead time, spatial/temporal resolution and accuracy of forecasts, and the specific parameters that should be reported in the forecast. Stakeholders should also be involved in activities such as vulnerability and risk analyses, since people at risk are more likely to take action on the basis of information they were involved in producing.

International-, national-, and local-level institutions all have important roles to play in fostering the development and implementation of a disease early warning system. For instance:

• *International/regional level*: Seasonal climate forecasts can be developed and/or disseminated by organizations such as the International Research Institute for Climate Prediction, NOAA's Climate Prediction Center, the European Centre for Medium-Range Weather Forecasting, and the World Meteorological Organization. Institutions such as the World Health Organization and the Pan American Health Organization can provide technical assistance and training to help develop national capabilities for surveillance and early warning systems.

• *National level:* Scientific teams working within national public health agencies or other organizations can use climate forecasts together with information from vulnerability assessments and surveillance networks to assess the disease risks posed to specific communities. National governments and scientific agencies can also help strengthen the basic infrastructure for epidemiological and environmental surveillance systems.

• *State/local level:* For areas determined to be at risk, disease watches/warnings can be disseminated to public health officials and other local authorities, who can then make decisions about intensifying surveillance or mobilizing intervention efforts. Local-level organizations, including community leaders and neighborhood representatives, can aid in activities such as monitoring vulnerability patterns, assessing intervention strategies and local response capacity, and strengthening networks for decision making and public communication.

Current Feasibility of Early Warning Systems

The previous sections describe the components of an effective climate-based disease early warning system. The feasibility of actually implementing such a system depends on numerous factors. For instance:

• It must be possible to provide sufficiently reliable climate forecasts for the region in question.
• There must be a strong understanding of the fundamental climate/disease linkages, so that the methods used to generate warnings offer a reasonable level of predictive value.
• There must be effective response measures available to implement within the window of lead time provided by the early warning.
• The community in question must be able to support the needed infrastructure such as surveillance systems and networks for disseminating information between organizations and to the general public.

Based on these criteria, there are at present very few contexts in which establishment of an effective, operational early warning system is entirely feasible. Exceptions may include cases such as the one described in Box 7-2, where a qualitative understanding of climate/disease associations is sufficient to warrant low-risk or "no-regrets" intervention actions. If progress is made in meeting the challenges listed above, however, there is *potential* for eventually developing effective disease early warning systems in many contexts. In the meantime, valuable information can be gained from further development of models to create "experimental" disease forecasts and from pilot projects that foster new scientific and institutional collaborations and provide real-world experience in using seasonal forecasts to meet the needs of specific communities. There have been several examples of such pilot programs in recent years, supported by institutions such as NOAA's Office of Global Programs, the International Research Institute for Climate Prediction, and the Inter-American Institute.

Case Study: West Nile Virus—Risk and Response

The emergence of West Nile encephalitis during the summer of 1999 in New York City, the first report of the disease in the United States, provides a useful example of several concepts discussed in this report. For instance, it illustrates the importance of interdisciplinary scientific approaches (in this case, the need for cooperation between scientists who deal with animal health and those that deal with human health). It also demonstrates the challenges of coordinating response actions among various government agencies and public health officials, and effectively communicating about a new disease risk to the public.

On August 23 an outbreak of human encephalitis was reported to the New York City Department of Health. On September 3 the CDC announced that the disease was the mosquito-borne St. Louis encephalitis (SLE). In response, New York City immediately initiated a campaign of aerial and ground spraying of pesticides to reduce the population of mosquitos. At this time the same area had reports of dead crows and viral encephalitis of unknown origin in several exotic bird species, including a Chilean flamingo at the Bronx Zoo. The SLE virus does not normally cause disease in birds; hence, at first the CDC assumed that the human and avian outbreaks were unrelated. On September 24, however, results presented from further immunological and molecular tests of samples taken from humans and birds revealed that both epidemics were caused by West Nile, a virus related to the SLE virus but unknown in the Western Hemisphere until that time (Briese et al., 1999; Jia et al., 1999; Lanciotti et al., 1999).

Risk and vulnerability. Mosquito-borne illness was extremely rare in New York City before the 1999 outbreak of West Nile virus. This contributed to the vulnerability of the population prior to the outbreak and the high perception of risk after the disease was publicized. Vulnerability came from the low emphasis on mosquito control, low behavioral avoidance of mosquitos (standing water, unscreened windows, outdoor evening activities), and the lack of immunity in the population. The perception of risk was high because of the relative novelty of a mosquito-borne disease, the lack of an effective treatment, and the potential seriousness of the symptoms including death. In 1999 there were 61 confirmed cases of West Nile virus in the New York City area, of which seven were fatal. Additionally, thousands of crows and other birds, nine horses, and one cat were reported to have died from the disease (Asnis et al., 2000).

West Nile virus was isolated in 1999 from *Culex pipiens* mosquitos, a species that is abundant in New York City. This mosquito feeds on both birds and people, usually in the evening. West Nile virus-infected adult *Cu. pipiens* were found during the winter, hidden in relatively warm, secluded shelters, which may explain why transmission of the virus continued during 2000. At least three species of *Aedes* mosquitos, which may bite during the daytime, also were found to be infected in and around New York City.

Preparedness and response. Health officials identified and responded to the outbreak of encephalitis relatively quickly. The disease was first misidentified because the symptoms and pathology of West Nile virus and SLE are similar and because the fluid and tissue samples from patients cross-reacted with SLE in the initial antibody screen. In this instance, however, the consequences of misidentification were minimal because the response measures for SLE and West Nile were the same. The New York City Department of Health was notified on August 23, 1999, of suspected cases of encephalitis, and mosquito control measures were activated after the cases were identified as SLE two weeks later. Although the response was rapid, it was still late relative to the peak of the epidemic.

To manage the potential spread of West Nile virus, the New Jersey Department of Health and Senior Services developed a preparedness plan that included the following components:

• Human, animal, and mosquito surveillance to detect whether and where the virus is present, including sentinel chicken flocks.

• Continued comprehensive mosquito control activities to suppress the emergence of mosquito populations.

• A coordinated system of reporting and laboratory testing of samples.

• A wide-reaching communication and education plan for the public and professionals (educational materials distributed through local health departments), including media releases, a Web site, and a strengthened alert network of physicians and hospital personnel.

In the summer of 2000, surveillance programs detected West Nile virus-infected birds in five northeastern states—New York, New Jersey, Massachusetts, Connecticut, and Maryland—indicating epizootic transmission of the disease through a much wider geographic area than in 1999.

Public communication. Why was the response to West Nile virus so extensive and crisis oriented? In the month after the first reported cases of encephalitis, the *New York Times* published 47 articles on the epidemic. In the following year, 30 articles were published during the month of July alone, indicating that

---

Box 7-3
Early Warning for Water Quality

The water industry has unique needs for early warning systems. Most developed countries have routine programs in place to monitor drinking water supplies for a variety of organic and inorganic chemicals, microbes (such as giardia lamblia, legionella, total coliforms, enteric viruses), and turbidity. In coastal waters some regions have monitoring for algal blooms to warn of toxins, and biological monitoring of fish is sometimes used as a catch-all for water safety. These types of monitoring systems can be useful for vulnerability assessments but are not much use for predictive purposes. Although it is known that runoff from heavy rainfall and snowmelt can impact some water quality systems (see discussion in Chapter 4), drinking water managers do not currently make any decisions based on rainfall or other climatic data.

Creating systems to provide early warning for the presence of water-borne pathogens will require improving our fundamental understanding of the linkages between climatic/ecological changes and waterborne disease outbreaks and developing hydrological models that can provide quantitative forecasts of risks in specific watersheds and water distribution systems. A more immediate concern, however, is the fact that many developing countries do not currently have the resources for maintaining even basic water treatment facilities or systems to monitor for the presence of water-borne pathogens. Without such basic infrastructure in place, the feasibility and effectiveness of any type of early warning system would be highly limited.

---

public interest had not diminished. The actual risk of contracting encephalitis was not communicated effectively relative to the intensity of the emergency response. In some cases people contacted the health department after finding a mosquito in their home or called a hospital emergency room after being bitten by one (*New York Times*, June 4, 2000). The pesticide-spraying campaign by itself caused alarm in the community, both from the implication that there was a health crisis and concern over side effects of heavy pesticide use. Another factor that elevated public concern was the novelty of the disease. Even prior to identification of the disease as West Nile, mosquito-borne viral encephalitis was an exotic disease for most New Yorkers, eclipsing more familiar diseases such as influenza that pose a much greater health threat.

Potential climate linkages. West Nile virus has been found in a broad range of climates, reflecting a broad mosquito host range. It has been previously described in Central Europe, the Middle East, and sub-Saharan Africa, with some evidence that it may be spread by bird migration between Europe and Africa (Watson et al., 1972; Miller et al., 2000). West Nile virus is endemic in temperate areas and was probably brought to New York through international transport of an infected bird, mosquito, or human. There has, however, been speculation that climatic factors in 1999, particularly the mild winter followed by an especially hot, dry summer, contributed to the outbreak (Epstein, 2000). Others have made the argument that West Nile virus represents the type of health threat that could increase with global warming, stating for instance that "this harrowing event in New York only foreshadows what more temperate-climate countries can expect unless we all work to reverse global warming." (Musil, 1999). It is certainly worthwhile to consider possible climatic influences on the emergence of West Nile encephalitis, but any claims about potential linkages should receive rigorous scientific review before they are used to make projections about future disease trends.

EXAMPLES OF THE USE OF CLIMATE FORECASTS

There are currently no examples of truly operational climate-based disease early warning systems. In some instances, investigators have begun developing models that use environmental observations to predict disease outbreaks (such as Rift Valley fever and hantavirus, discussed earlier), but thus far these models have only been used to make such "predictions" retroactively. Other realms of societal activity, however, do routinely use environmental observations and climate forecasts to issue early warnings to protect public welfare and safety and for resource and economic planning. We review here several of these examples—in the areas of agriculture, famine warning, forest fire control, and hurricane preparedness—for the purpose of highlighting useful analogies and lessons for the development of disease early warning systems.

ENSO Forecasts and Agricultural Planning

A number of studies have documented the value of ENSO forecasts for agriculture in the United States (e.g., Adams et al., 1995; Mjelde et al., 1997; Solow et al., 1998; Weiher, 1999), and these forecasts are used for making cropping decisions in Australian wheat production. Given Australia's location close to the "center of action" of ENSO events (i.e., the tropical Pacific) and the high climatic variability it experiences, its use of seasonal forecasting for agricultural production has been relatively well developed compared to other countries. Several decision-making tools have been developed, such as the RAINMAN and WHEATMAN software programs (Woodruff, 1992), which can assist farmers in making management decisions based on rainfall forecasts. The fundamental assumption in this research is that forecasts have value only if they can change decisions in a way that improves outcomes; in this context the question becomes: Can existing skill in seasonal rainfall prediction improve profitability by tactical modifications of management decisions?

Hammer et al. (1996) investigated the value of seasonal rainfall forecasts for making improved decisions regarding planting dates, varietal characteristics, and application of nitrogen fertilizer. The assumed strategy is that the forecast must generate a preferred combination of profit and risk compared to the fixed strategy (based on no forecast). They performed a cost-benefit analysis, determining profit as a function of wheat production, price, and variable and fixed costs. They also determined the average profit for each nitrogen application rate over all years and the average profit based on each phase of the ENSO cycle (Stone et al., 1996). Comparing results of the fixed strategy and the tactical strategy (considering the forecast ENSO phase), it was found that up to a 20 percent increase in profit and 35 percent reduction in risk could result by using forecast information.

While it is useful to determine the theoretical value of forecasts, the more important question is whether the forecasts are actually being used. In the case of the wheat farmers, the answer appears to be yes, as evidenced by the fact that the Queensland seasonal forecast Web site receives 38,000 hits per annum, and the Queensland bureau receives 1,200 fax requests for the forecast per annum, many of which are from farmers.

An important concept that this work has highlighted is that a forecast in any given year is not guaranteed; rather, forecasting essentially just shifts the probabilities of outcomes. The value of forecasts to farmers thus requires an appreciation of the concept of probability. In the case of the 1998 wheat crop, expected low yields failed to materialize in many cases despite strong El Niño conditions. Essentially in those regions of northeast Australia that had high soil moisture at the time of planting, yields close to normal were obtained, whereas the areas where initial soil moisture was low, lower than normal yields were obtained. Also, contrary to typical conditions for an El Niño year, rains did occur during the flowering phase (Meinke and Hochman, 2000). This reflects the complexity of the wheat cropping conditions and how these can interact with seasonal rainfall. Dealing with the problem of false positives for decision making is one of the central challenges in the use of seasonal forecasts in agriculture.

In a comprehensive review of the use of seasonal forecasts in agriculture in Australia (Hammer, 2000), five significant lessons emerged:

• Understanding and predicting responses of the agricultural system are critical.
• Application of seasonal forecasts concerns managing risks.
• Forecasting information must be directly relevant to the decision makers.
• Considerable effort must be expended to appropriately communicate probabilistic information.
• There must be further consideration of connecting agricultural and climate models.

Forest Fire Control

Short-term weather forecasts have been used for fire prediction systems in the United States since the early 1950s. The system primarily used today—the National Fire Danger Rating System—issues fire danger warnings based on once-daily weather observations and indices such as dead fuel moisture, live fuel moisture (greenness maps based on NDVI), drought maps, lower-atmospheric stability index maps, and lightening ignition efficiency values. These warnings are used to make decisions about prohibiting the building of open fires, shutting down logging activities, and allocating resources such as manpower and machinery.

Short-term weather forecasts are used in the course of a forest fire, and resource dispatch centers do sometimes use medium term forecasts (10-30 days) to determine what kinds of resources to have on standby. Given the short-term nature of the decision-making structure for controlling forest fires, there currently is little operational use for seasonal-term climate forecasts, but there may be some potential for this if the U.S. Forest Service was convinced of the forecasting predictive power, especially given the fact that there is a well-defined forest fire "season" in most regions. There is also some evidence tying the ENSO cycle with the frequency and intensity of forest fires worldwide (WMO, 1999; National Research Council, 1998). The U.S. Department of Agriculture's Forest Service/Forest Fire Laboratory prepares monthly fire weather forecasts (McCutchan et al., 1991) and a monthly "fire potential" calculated from temperature and relative humidity forecasts. These measures, however, are considered to be too spatially coarse, and their use is hampered by the generally poor accuracy of forecasts made at the monthly time scale. There has been some research on producing longer-term fire severity forecasts, but this work is still very preliminary (Roads et al., 2000).

Some of the lessons that fire managers have learned about how people actually use forecasts, which are likely applicable to a disease early warning system, include the following:

- Even with sophisticated models and other forecasting tools at their disposal, resource managers often rely primarily on expert judgment to make decisions. Likewise, personal judgment is still required to make decisions about what constitutes an acceptable risk.
- Systems that do rely primarily on expert judgment can be left in a vulnerable state if there is a shortage of people with sufficient experience (e.g., when experienced managers retire).
- In systems where several different groups are drawing on a common limited pool of resources for mobilizing preparedness and response actions, a forecast of impending risk can lead to considerable organizational tensions resulting from competition for resources.

Hurricane Forecasting

Hurricanes are among the weather entities most destructive to human society. They can cause enormous economic loses (e.g., Hurricane Andrew in 1992), and in countries with highly vulnerable populations the effects can be far more devastating with enormous loss of life (such as occurred with Hurricane Mitch in 1998).

There is a relatively well-developed prediction system for hurricanes operated by the U.S. National Hurricane Prediction Center in Miami, Florida. The two main aspects of a hurricane that are forecast are its track and intensity, which are forecast for 24-, 48-, and 72-hour periods using suites of dynamic and empirical models. While dynamical models and statistical techniques have developed significantly over the past 20 years, significant forecasting errors still occur. As of 1997, average landfall errors were about 115 miles (for 24-hour forecasts), an improvement of only about 20 miles from 20 years earlier (Pielke and Pielke, 1997).

It is these relatively short-term forecasts that are most critical for decision making regarding protection of life and property; however, longer-range hurricane forecasts are also produced (e.g., Gray et al., 1999). These forecasts are made in early August for the tropical cyclone season of that year (August through October). Forecasts are also made based on the state of the ENSO cycle, but these are very general forecasts that are not yet detailed enough to be of much use in decision making.

The main effects of hurricanes that reach land result from some combination of storm surges, winds, and rainfall. One of the societal problems associated with these events is that a high percentage of residents of coastal areas vulnerable to hurricanes have no personal understanding of the possible magnitudes of damage (Pulwarty and Riebsame, 1997), which affects the way they respond to hurricane warnings.

The main decisions related to hurricane prediction include issuing watches and warnings, issuing evacuation plans (and safety advice for those who do not leave), and protection of property through specific building codes and land use decisions. The evacuations issued in 1999 during Hurricane Floyd for the coast of South Carolina are a good example of the problem of false positives, since the hurricane did not make landfall where forecast, and of inadequate evacuation planning, which led to massive traffic congestion.

Some of the lessons learned from forecasting and responding to hurricanes that might be applicable to a disease early warning system include:

- the need to prepare the population for the possibility of false-positive forecasts;
- the challenge of educating populations that have not previously been exposed to the hazard in question; and
- the need to assess the vulnerability of different populations within one region.

Famine Early Warning System

The Famine Early Warning System (FEWS), operated by the U.S. Agency for International Development, is an information system designed to help decision makers prevent famine in sub-Saharan Africa by allowing them to better understand the causes of famine, detect changes that create famine risks, and determine appropriate famine mitigation and prevention strategies. FEWS focuses on monitoring high-risk countries where populations are particularly vulnerable to episodic food shortages. A wide variety of information is used to develop forecasts of the "food security" of these populations, including remote sensing (e.g., normalized difference vegetation index), rainfall, crop growth, crop production, and demographic, socioeconomic, and health information. FEWS provides regional and country-specific analyses, monthly bulletins integrating forecast information, and vulnerability assessments. These assessments are typically carried out in conjunction with local institutions, and the analyses and conclusions are shared with the host governments. The main steps of this system are summarized in Box 7-4.

Box 7-4
Steps in FEWS Early Warning

Pre-season vulnerability assessment: analysis conducted prior to the growing season to identify populations likely to be hard-hit in the case of drought or other shock.

Season monitoring: analysis of satellite imagery on rainfall and crop growth and reporting of cereal price data, produced by a number of different groups and collated by FEWS.

Special alerts and warnings: briefings, cables, and e-mails to inform of potential food emergencies.

Contingency planning: efforts undertaken during poor crop production years to monitor food security situations and determine appropriate responses using the following monitoring instruments: preliminary harvest assessments, cereal shortfall estimates, needy population targeting, and food needs assessment.

Food aid intervention evaluation: assessments to understand targeting methods used by nongovernmental organizations, to gain insight into the nature of vulnerability and to observe community status after intervention.

SOURCE: Famine Early Warning System, 1999.

Some of the difficulties that the operators of the system have encountered include a lack of clarity about main uses and users of data, lack of coordination and structure in data collection systems, and an unclear relationship between short-term crisis intervention and activities designed to address longer-term underlying problems. Some of the critical lessons learned include the following:

• Involvement of the host country in developing the system and disseminating information is critical. Otherwise, the local population will not trust the information and it will not be used.

• The methods used must be made clear to the local population, since lack of transparency in the system also leads to distrust and lack of use.

• Even when sophisticated forecasts and modeling tools are available, people will often rely on simple intuitive information to make decisions (e.g., Is it raining?).

• Perhaps most importantly, an early warning system is of little value if it is not connected to a system for making decisions about possible responses.

FEWS offers perhaps the most relevant example of an early warning system for the public health community to consider. In fact, since malnutrition and susceptibility to disease are closely linked, it could be highly advantageous for a disease early warning system to build on the infrastructure and personal networks that FEWS has already established in many regions.

8

Key Findings and Recommendations

In this chapter we highlight the key findings related to each aspect of the committee's charge and recommend priorities for future research.

KEY FINDINGS: LINKAGES BETWEEN CLIMATE AND INFECTIOUS DISEASES

Weather fluctuations and seasonal-to-interannual climate variability influence many infectious diseases. The characteristic geographic distributions and seasonal variations of many infectious diseases are *prima facie* evidence of linkages with weather and climate. Studies have shown that factors such as temperature, precipitation, and humidity affect the lifecycle of many disease pathogens and vectors (both directly, and indirectly through ecological changes) and thus can potentially affect the timing and intensity of disease outbreaks. However, disease incidence is also affected by factors such as sanitation and public health services, population density and demographics, land use changes, and travel patterns. The importance of climate relative to these other variables must be evaluated in the context of each situation.

Observational and modeling studies must be interpreted cautiously. There have been numerous studies showing an association between climatic variations and disease incidence, but such studies are not able to fully account for the complex web of causation that underlies disease dynamics and thus may not be reliable indicators of future changes. Likewise, a variety of models have been developed to simulate the effects of climatic changes on incidence of diseases such as malaria, dengue, and cholera. These models are useful heuristic tools for testing hypotheses and carrying out sensitivity analyses, but they are not necessarily intended to serve as predictive tools, and often do not include processes such as physical/biological feedbacks and human adaptation. Caution must be exercised then in using these models to create scenarios of future disease incidence, and to provide a basis for early warnings and policy decisions.

The potential disease impacts of global climate change remain highly uncertain. Changes in regional climate patterns caused by long-term global warming could affect the potential geographic range of many infectious diseases. However, if the climate of some regions becomes more suitable for transmission of disease agents, human behavioral adaptations and public health interventions could serve to mitigate many adverse impacts. Basic public health protections such as adequate housing and sanitation, as well as new vaccines and drugs, may limit the future distribution and impact of some infectious diseases, regardless of climate-associated changes. These protections, however, depend upon maintaining strong public health programs and assuring vaccine and drug access in the poorer countries of the world.

Climate change may affect the evolution and emergence of infectious diseases. Another important but highly uncertain risk of climate change are the potential impacts on the evolution and emergence of infectious disease agents. Ecosystem instabilities brought about by climate change and concurrent stresses such as land use changes, species dislocation, and increasing global travel could potentially influence the genetics of pathogenic microbes through mutation and horizontal gene transfer, and could give rise to new interactions among hosts and disease agents. Such changes may foster the emergence of new infectious disease threats.

There are potential pitfalls in extrapolating climate and disease relationships from one spatial/temporal scale to another. The relationships between climate and infectious disease are often highly dependent upon local-scale parameters, and it is not always possible to extrapolate these relationships meaningfully to broader spatial scales. Likewise, disease impacts of seasonal to interannual climate variability may not always provide a useful analog for the impacts of long-term climate change. Ecological responses on the timescale of an El Niño event, for example, may be significantly different from the ecological responses and social adaptations expected under long-term climate change. Also, long-term climate change may influence regional climate variability patterns, hence limiting the predictive power of current observations.

Recent technological advances will aid efforts to improve modeling of infectious disease epidemiology. Rapid advances being made in several disparate scientific disciplines may spawn radically new techniques for modeling of infectious disease epidemiology. These include advances in sequencing of microbial genes, satellite-based remote sensing of ecological conditions, the development of Geographic Information System (GIS) analytical techniques, and increases in inexpensive computational power. Such technologies will make it possible to analyze the evolution and distribution of microbes and their relationship to different ecological niches, and may dramatically improve our abilities to quantify the disease impacts of climatic and ecological changes.

KEY FINDINGS: THE POTENTIAL FOR DISEASE EARLY WARNING SYSTEMS

As our understanding of climate/disease linkages is strengthened, epidemic control strategies should aim towards complementing "surveillance and response" with "prediction and prevention." Current strategies for controlling infectious disease epidemics depend largely on surveillance for new outbreaks followed by a rapid response to control the epidemic. In some contexts, however, climate forecasts and environmental observations could potentially be used to identify areas at high risk for disease outbreaks and thus aid efforts to limit the extent of epidemics or even prevent them from occurring. Operational disease early warning systems are not yet generally feasible, due to our limited understanding of most climate/disease relationships and limited climate forecasting capabilities. But establishing this goal will help foster the needed analytical, observational, and computational developments.

The potential effectiveness of disease early warning systems will depend upon the context in which they are used. In cases where there are relatively simple, low-cost strategies available for mitigating risk of epidemics, it may be feasible to establish early warning systems based only on a general understanding of climate/disease associations. But in cases where the costs of mitigation actions are significant, a precise and accurate prediction may be necessary, requiring a more thorough mechanistic understanding of underlying climate/disease relationships. Also, the accuracy and value of climate forecasts will vary significantly depending on the disease agent and the locale. For instance, it will be possible to issue sufficiently reliable ENSO-related disease warnings only in regions where there are clear, consistent ENSO-related climate anomalies. Finally, investment in sophisticated warning systems will be an effective use of resources only if a country has the capacity to take meaningful actions in response to such warnings, and if the population is significantly vulnerable to the hazards being forecast.

Disease early warning systems cannot be based on climate forecasts alone. Climate forecasts must be complemented by an appropriate suite of indicators from ongoing meteorological, ecological, and epidemiological surveillance systems. Together, this information could be used to issue a "watch" for regions at risk and subsequent "warnings" as surveillance data confirm earlier projections. Development of disease early warning systems should also include vulnerability and risk analysis, feasible response plans, and strategies for effective public communication. Climate-based early warning systems being developed for other applications, such as agricultural planning and famine prevention, provide many useful lessons for the development of disease early warning systems.

Development of early warning systems should involve active participation of the system's end users. The input of stakeholders such as public health officials and local policymakers is needed in the development of disease early warning systems, to help ensure that forecast information is provided in a useful manner and that effective response measures are developed. The probabilistic nature of climate forecasts must be clearly explained to the communities using these forecasts, so that response plans can be developed with realistic expectations for the range of possible outcomes.

RECOMMENDATIONS FOR FUTURE RESEARCH AND SURVEILLANCE

Research on the linkages between climate and infectious diseases must be strengthened. In most cases, these linkages are poorly understood and research to understand the causal relationships is in its infancy. Methodologically rigorous studies and analyses will likely improve our nascent understanding of these linkages and provide a stronger scientific foundation for predicting future changes. This can best be accomplished with investigations that utilize a variety of analytical methods (including analysis of observational data, experimental manipulation studies, and computational modeling), and that examine the consistency of climate/disease relationships in different societal contexts and across a variety of temporal and spatial scales. Progress in defining climate and infectious disease linkages can be greatly aided by focused efforts to apply recent technological advances such as remote sensing of ecological changes, high-speed computational modeling, and molecular techniques to track the geographic distribution and transport of specific pathogens.

Further development of disease transmission models is needed to assess the risks posed by climatic and ecological changes. The most appropriate modeling tools for studying climate/disease linkages depend upon the scientific information available. In cases where there is limited understanding of the ecology and transmission biology of a particular disease, but sufficient historical data on disease incidence and related factors, statistical-empirical models may be most useful. In cases where there are insufficient surveillance data, "first principle" mechanistic models that can integrate existing knowledge about climate/disease linkages may have the most heuristic value. Models that have useful predictive value will likely need to incorporate elements of both these approaches. Integrated assessment models can be especially useful for studying the relationships among the multiple variables that contribute to disease outbreaks, for looking at long-term trends, and for identifying gaps in our understanding.

Epidemiological surveillance programs should be strengthened. The lack of high-quality epidemiological data for most diseases is a serious obstacle to improving our understanding of climate and disease linkages. These data are necessary to establish an empirical basis for assessing climate influences, for establishing a baseline against which one can detect anomalous changes, and for developing and validating models. A concerted effort, in the United States and internationally, should be made to collect long-term, spatially resolved disease surveillance data, along with the appropriate suite of meteorological and ecological observations. Centralized, electronic databases should be developed to facilitate rapid, standardized reporting and sharing of epidemiological data among researchers.

Observational, experimental, and modeling activities are all highly interdependent and must progress in a coordinated fashion. Experimental and observational studies provide data necessary for the development and testing of models; and in turn, models can provide guidance on what types of data are most needed to further our understanding. The committee encourages the establishment of research centers dedicated to fostering meaningful interaction among the scientists involved in these different research activities through long-term collaborative studies, short-term information-sharing projects, and interdisciplinary training programs. The National Center for Ecological Analysis and Synthesis provides a good model for the type of institution that would be most useful in this context.

Research on climate and infectious disease linkages inherently requires interdisciplinary collaboration. Studies that consider the disease host, the disease agent, the environment, and society as an interactive system will require more interdisciplinary collaboration among climate modelers, meteorologists, ecologists, social scientists, and a wide array of medical and public health professionals. Encouraging such efforts requires strengthening the infrastructure within universities and funding agencies for supporting interdisciplinary research and scientific training. In addition, educational programs in the medical and public health fields need to include interdisciplinary programs that explore the environmental and socioeconomic factors underlying the incidence of infectious diseases.

Numerous U.S. federal agencies have important roles to play in furthering our understanding of the linkages among climate, ecosystems, and infectious disease. There have been a few programs established in recent years to foster interdisciplinary work in applying remote-sensing and GIS technologies to epidemiological investigations. The committee applauds these efforts and encourages all of the relevant federal agencies to support interdisciplinary research programs on climate and infectious disease, along with an interagency working group to help ensure effective coordination among these different programs. The U.S. Global Change Research Program (USGCRP) may provide an appropriate forum for this type of coordinating body. This will require, however, that organizations such as the Centers for Disease Control and Prevention, and the National Institute of Allergy and Infectious Diseases become actively involved with the USGCRP.

In closing, the committee wishes to emphasize that even if we are able to develop a strong understanding of the linkages among climate, ecosystems, and infectious diseases, and in turn, are able to create effective disease early warning systems, there will always be some element of unpredictability in climate variations and infectious disease outbreaks. Therefore, a prudent strategy is to set a high priority on reducing people's overall vulnerability to infectious disease through strong public health measures such as vector control efforts, water treatment systems, and vaccination programs.

Acronyms/Abbreviations

AGCC	Anthropogenic Global Climate Change
AVHRR	Advanced Very High Resolution Radiometer
CDC	Centers for Disease Control and Prevention
CHAART	Center for Health Applications of Aerospace Related Technologies
DoD	Department of Defense
ENSO	El Niño/Southern Oscillation
FEWS	Famine Early Warning System
GCM	General Circulation Models
GIS	Geographic Information System
HPS	Hantavirus Pulmonary Syndrome
IA	Integrated Assessment
IPCC	Intergovernmental Panel on Climate Change
NDVI	Normalized Difference Vegetation Index
NOAA	National Oceanic and Atmospheric Administration
NRC	National Research Council
PAHO	Pan American Health Organization
RVF	Rift Valley Fever
SEIR	Susceptible, Exposed, Infected, Removed
SLE	St. Louis Encephalitis
USGCRP	United States Global Change Research Program
WHO	World Health Organization
WMO	World Meteorological Organization

Glossary

This glossary has been compiled to help familiarize readers with some basic terminology used in such disciplines as climatology, ecology, medicine, and public health. It is not meant to be a comprehensive or definitive glossary for any of these fields.

Agent (of disease): Factor such as a microorganism whose presence is essential for the occurrence of a disease.

Albedo: Measure of the reflecting power of an object (e.g., Earth), expressed as the proportion of incident light it reflects.

Anthropogenic: Caused or produced by humans.

Anthroponosis: Infection that causes disease in nonhuman hosts but that can be perpetuated solely in human hosts (e.g., malaria).

Association: Statistical dependence between two or more events, characteristics, or other variables. The presence of an association does not necessarily imply a causal relationship.

Case fatality ratio: Cumulative incidence of death in a group of individuals who develop a particular disease over a specific time period.

Climate: Average meteorological conditions over a specified time period, usually at least a month, resulting from interactions among the atmosphere, oceans, and land surface. Climate variations occur over a wide range of spatial and temporal scales.

Disease, communicable: Illness caused by a specific infectious agent; arises through transmission of that agent or its products from an infected person, animal, or reservoir to a susceptible host.

Dose-response relationship: Relationship in which a change in the amount, intensity, or duration of exposure is associated with a change in the risk of a specified outcome.

Ecosystem: Mutually interrelated communities of species and abiotic components, existing as a system with specific interactions and exchange of matter, energy, and information.

El Niño: A warming of the surface waters of the tropical Pacific that occurs every three to five years, temporarily affecting weather worldwide.

Endemic: Restricted or peculiar to a locality or region. Endemic infection refers to a sustained, relatively stable pattern of infection in a specified population.

Epidemic: Appearance of an abnormally high number of cases of infection in a given population.

Epidemiology: Study of the distribution and determinants of health-related states or events in specified populations. Epidemiology is the basic quantitative science of public health.

Epizootic: An epidemic in an animal host population.

Extrinsic incubation period: Time required for the development of a disease agent in a vector from the time of uptake of the agent to the time when the vector is infective.

Herd immunity: Mechanism by which an infection may be eradicated from a population, although some susceptibles still remain, because the remainder of the population is immune and thus transmission is reduced.

Host (disease): Person or other living animal that affords subsistence or lodgment to an infectious agent under natural conditions.

Immunity: Condition of protection against infectious disease conferred either by the immune response generated by immunization or previous infection, or by other non-immunological factors.

Incidence: Number of cases of a disease commencing, or of persons falling ill, during a given period of time in a specified population. Incidence rate is the number of new cases of a specific disease diagnosed or reported during a defined interval of time divided by the number of all persons in a defined population during the same time.

Infection: Presence of a parasite in a host, where it may or may not cause disease.

Infection, reservoir of: Any person, animal, arthropod, plant, soil or combination of these in which an infectious agent normally lives and multiplies, on which it depends primarily for survival, and where it reproduces itself in such a manner that it can be transmitted to a susceptible host.

Infectivity: Characteristic of the disease agent that embodies the capability to enter, survive, and multiply in the human host.

Meta-analysis: Process of using statistical methods to combine the results of different studies.

Miasma theory: An explanation for the origin of epidemics, based on the notion that when the air was of a so-called bad quality, persons breathing it would become ill.

Microclimate: In climatology, defined as localized climate, incorporating physical processes in the lowest 100 to 2,000 meters of the atmosphere. In ecology, defined as climatic conditions in the environmental space occupied by a species, a community of species, or an ecosystem.

Mitigation: Initiatives that reduce the risk from natural and man-made hazards. With respect to climate change, mitigation usually refers to actions taken to reduce the emissions or enhance the sinks of greenhouse gases.

Model, mathematical: Representation of a system, process, or relationship in mathematical form in which equations are used to simulate the behavior of the system or process under study.

Morbidity: State of ill health produced by a disease; rate of occurrence of disease in a population.

Outbreak: Localized occurrence as opposed to a generalized epidemic.

Pandemic: Epidemic that occurs over a very wide area.

Pathogen: Organism capable of causing disease.

Pathogenicity: Frequency with which an organism produces disease. Pathogenicity of an infectious agent is measured by the ratio of the number of persons developing clinical illness to the number exposed to infection.

Prevalence: Proportion of persons in a population that is currently affected by a particular disease. Prevalence rate is the number of cases of a specific disease at a particular point in time divided by the population at that time living in the same region.

Risk: Probability that an event will occur; a measure of the degree of loss expected by the occurrence of an event.

Risk assessment: Qualitative or quantitative estimation of the likelihood of adverse effects that may result from exposure to specific health hazards.

Scenario building: Method of predicting the future that relies on a series of assumptions about alternative possibilities, rather than on simple extrapolation or existing trends.

Southern Oscillation: A large-scale atmospheric and hydrospheric fluctuation centered in the equatorial Pacific Ocean; exhibits a nearly annual pressure anomaly, alternatively high over the Indian Ocean and high over the South Pacific; the variation in pressure is accompanied by variations in wind strengths, ocean currents, sea surface temperatures, and precipitation in the surrounding areas.

Susceptibility: Probability that an individual or population will be affected by an external hazard, such as infection by a pathogen.

Teleconnections: Statistical relationship between the El Niño/Southern Oscillation (ENSO) cycle and rainfall or temperature anomalies observed in a particular geographical location.

Transmission: Process by which a pathogen passes from a source of infection to a new host.

Vector: An organism, such as an insect, that transmits a pathogen from one host to another.

Vectorial capacity: Average number of potentially infective bites transmitted by one vector species from one infective host in one day.

Virulence: Degree of pathogenicity; disease-evoking power of a microorganism in a given host. Numerically, the ratio of the number of cases of overt disease in the total number infected.

Vulnerability: Extent to which a population is liable to be harmed by a hazard event. Depends on the population's exposure to the hazard and its capacity to adapt or otherwise mitigate adverse impacts.

Weather: Condition of the atmosphere at a particular place and time measured in terms of wind, temperature, humidity, atmospheric pressure, cloudiness, and precipitation. In most places, weather can change from hour to hour, from day to day, and from season to season.

Zoonosis: Infection that causes disease in human populations but that can be perpetuated solely in nonhuman host animals (e.g., bubonic plague).

References

Adams, R.M., K.J. Bryant, and R. Weiher, 1995. Value of improved long-range weather information. *Contemporary Economic Policy* 13:10-19.

Alterholt, T.B., M.W. LeChevalier, W.D. Norton, and J.S. Rosen. 1998. Effect of rainfall on giardia and crypto. *J. Amer. Water Works Assoc.* 90:66-80.

Anderson, R.M. and R.M. May. 1992. *Infectious Diseases of Humans: Dynamics and Control.* Oxford: Science Publications.

Ashford, O.M. 1985. *Prophet or Professor: The Life and Work of Lewis Fry Richardson.* Bristol and Boston: A. Hilger.

Asnis D.S., R. Conetta, A.A. Teixeira, G. Waldman, and B.A. Sampson. 2000. The West Nile Virus Outbreak of 1999 in New York: the Flushing Hospital experience. *Clin. Infect. Dis.* 30:413-418.

Aspray, W. 1990. *John von Neumann and the Origins of Modern Computing.* Cambridge, Mass.: MIT Press.

Baker, M.N., 1948. The quest for pure water. *J. Amer. Water Works Assoc.* Vol 1., second edition.

Barnston, A.G., A. Leetmaa, V.E. Kousky, R.E. Livezey, E.A. O'Lenic, H. van den Dool, A.J. Wagner and D.A. Unger. 1999. NCEP forecasts for the El Niño of 1997-98 and its U.S. impacts. *Bull. Amer. Meteor. Soc.* 80:1829-1852.

Barzun, J. 2000. *From Dawn to Decadence.* New York: Harper-Collins.

Bastien, J.W. 1998. *The Kiss of Death: Chagas' Disease in the Americas.* Salt Lake City: University of Utah Press.

Beck, L.R. 2000. Remote sensing and human health: New sensors and new opportunities. *Emerging Infectious Diseases* 6(3):217-226.

Bertrand, M.R., and M.L. Wilson. 1996. Microclimate-dependent survival of adult Ixodes scapularis (Acari: Ixodidae) in nature: life cycle and study design implications. *J. Med. Entomol.* 33:619-627.

Bissell, R. 1983. Delayed-impact infectious disease after a natural disaster. *J. Emergency Medicine* 1:59-66.

Bouma, M.J., H.E. Sondorp, and H.J. van der Kaay. 1994. Climate change and periodic epidemic malaria. *Lancet* 343:1440.

Bouma, M.J., and H.J. van der Kaay. 1996. The El Niño/Southern Oscillation and the historic malaria epidemics on the Indian subcontinent and Sri Lanka: an early warning system for future epidemics? *Trop. Med. Int. Health.* 1:86-96.

Bouma, M.J., and C. Dye. 1997. Cycles of malaria associated with El Niño in Venezuela. *J. Am. Med. Assoc.* 278:1772-1774.

Bouma, M.J., G. Poveda, W. Rojas, D. Chavasse, M. Quiñones, J. Cox, and J. Patz, 1997. Predicting high-risk years for malaria in Colombia using parameters of El Niño Southern Oscillation. *Trop. Med. Int. Health.* 2:1122-27.

Bradley D.J. 1993. Human tropical diseases in a changing environment, p.146-162. In: *Environmental Change and Human Health.* J. Lake, G. Brock, K. Ackrill, eds. Ciba Foundation Symposium, London, UK: CIBA Foundation.

Briese T., X.Y. Jia, C. Huang, L.J. Grady, and W.I. Lipkin. 1999. Identification of a Kunjin/West Nile-like flavivirus in brains of patients with New York encephalitis. *Lancet* 354:1261-1262.

Brown, J., S. Li and N. Bhagabati. 1999. Long-term trend toward earlier breeding in an American bird: a response to global warming? *Proc. Natl. Academy of Sciences* 96:5565.

Campbell, I., and J. McAndrews. 1993. Forest disequilibrium caused by rapid Little Ice Age cooling. *Nature* 366:336-338.

Carmichael, A.G. 1991. Contagion theory and contagion practice in fifteenth-century Milan. *Renaissance Quarterly.* 44:213-56.

Cassedy, J.H. 1986. *Medicine and American Growth.* Madison: University of Wisconsin Press.

Chan, N.Y., K.L. Ebi, F. Smith, T.F Wilson, and A.E. Smith. 1999. An integrated assessment framework for climate change and infectious diseases. *Environmental Health Perspectives* 107: 329-337.

Childs, J.E, T.G. Ksiazek, C.F. Spiropoulou, J.W. Krebs, S. Morzunov, G.O. Maupin, K.L. Gage, P.E. Rollin, J. Sarisky, and R.E. Enscore. 1994. Serologic and genetic identification of *Peromyscus maniculatus* as the primary rodent reservoir for a new hantavirus in the southwestern United States. *J. Infect. Dis.* 169:1271-1280.

Cipolla, C.M. 1981. *Fighting the Plague in Seventeenth-Century Italy.* Madison: University of Wisconsin Press.

Coleman, W. 1987. Koch's comma bacillus: The first year. *Bulletin of the History of Medicine* 61:315-342.

Colwell, R.R. 1996. Global climate and infectious diseases: The cholera paradigm. *Science* 274:2025-2031.

Colwell, R.R., and D.J. Grimes, eds. 2000. *Nonculturable Microorganisms in the Environment.* Washington D.C.: ASM Press. 354 pp.

Corradi, A., 1972. *Annali delle epidemie occorse in Italia, dalle prime memorie fino al 1850.* Five volumes. Bologna: Forni.

Creighton, C. 1969. *A History of Epidemics in the British Isles.* Two volumes. Cambridge: Cambridge University Press.

Crick, H., and T. Sparks. 1999. Climate change related to egg-laying trends. *Nature* 399:423-424.

Daubney, R., J.R. Hudson, and P.C. Garnham. 1931. Enzootic hepatitis or Rift Valley fever: An undescribed disease of sheep, cattle and man from East Africa. *J. Pathol. Bacteriol.* 89:545-579.

Davies, F.G., K.J. Linthicum, and A.D. James. 1985. Rainfall and epizootic Rift Valley fever. *Bulletin of the World Health Organization* 63:941-963.

Davis, M., and C. Zabinski. 1992. Changes in geographical range resulting from greenhouse warming: Effects of biodiversity in forests, pp. 297-308 in *Global Warming and Biological Diversity*, R. Peters and T. Lovejoy, eds. New Haven, Conn.:Yale University Press.

Dennis, D.T. 1998. Epidemiology, Ecology, and Prevention of Lyme Disease. in D.W. Rahn and J. Evans (eds). *Lyme Disease.* Philadelphia: American College of Physicians.

Desaive, J-P. 1972. *Médecins, climat et épidemies à la fin du XVIIIe siècle.* Paris and the Hague: Mouton & Co.

Dunne, J. 2000. Climate Change Impacts on Community and Ecosystem Properties: Integrating Manipulations and Gradient Studies in Montane Meadows. Ph.D. Dissertation, University of California, Berkeley.

Eisenberg, J.N., E.Y.W. Seto, A.W. Olivieri, and R.C Spear. 1996. Quantifying water pathogen risk in an epidemiological framework. *Risk Anal.* 16:549-563.

Epstein, P.E. 2000. Is global warning harmful to human health? *Scientific American* August:50-57.

Evans, A.S. 1978. Causation and disease: a chronological journey. *American Journal of Epidemiology* 108:249-258.

Ewald, P.W. 1994. *Evolution of Infectious Disease.* New York: Oxford University Press.

Famine Early Warning System. 1999. *FEWS Current Vunerability Assessment Guidance Manual.* U.S. Agency for International Development, Washington, D.C.

Farmer, P., 1999. *Infections and Inequality: The Modern Plagues.* Berkeley: University of California Press.

Fayer, R., ed. 1997. *Cryptosporidium and Cryptosporidiosis.* Boca Raton, FL: CRC Press.

Fein, J.S. and P.L. Stephens, (eds.) 1987. *Monsoons.* New York: John Wiley & Sons.

Fleming, J.R. 1990. *Meteorology in America, 1800-1870.* Baltimore: Johns Hopkins University Press.

Flynn N.M., P.D. Hoeprich, M.M. Kawachi, K.K. Lee, R.M. Lawrence, E. Goldstein, G.W. Jordan, R.S. Kundargi, and G.A. Wong. 1979. An unusual outbreak of windborne coccidioidomycosis. *New England Journal of Medicine* 301(7):358-361.

Focks D.A., E. Daniels, D.G. Haile, and J.E. Keesling. 1995. A simulation model of the epidemiology of urban dengue fever: Literature analysis, model development, preliminary validation, and samples of simulation results. *Am. J. Trop. Med. Hyg.* 53:489-506.

Focks D.A., R.A. Brenner, E. Daniels, and J. Hayes. 2000. Transmission thresholds for dengue in terms of Aedes aegypti pupae per person with discussion of their utility in source reduction efforts. *Am. J. Trop. Med. Hyg.* 62:11-18.

Freimuth, V., H.W. Linnan, and P. Potter. 2000. Communicating the threat of emerging infections to the public. *Emerging Infectious Diseases* 6(4):337-346.

Friedman, L.N, M.T. Williams, T.P. Singh, and T.R. Frieden. 1996. Tuberculosis, AIDS, and death among substance abusers on welfare in New York City. *New England Journal of Medicine* 334: 828-833.

Friedman, R.M. 1989. *Appropriating the Weather: Vilhelm Bjerknes and the Construction of a Modern Meteorology.* Ithaca, N.Y.: Cornell University Press.

Frisinger, H.H. 1977. *The History of Meteorology to 1800.* American Meteorological Society: Historical monograph series. New York: Science History Publications.

Garbe, P.L., B.J. Davis, J.S. Weisfeld, L. Markowitz, P. Miner, F. Garrity, J.M Barbaree, and A.L. Reingold. 1985. Nosocomial Legionnaires' disease: epidemiologic demonstration of cooling towers as a source. *J. Am. Med. Assoc.* 254:521-524.

Giacomini, T., O. Axler, J. Mouchet, and P. Lebrin. 1997. Pitfalls in the diagnosis of airport malaria: Seven cases observed in the Paris area in 1994. *Scandinavian Journal of Infectious Diseases* 29:433-435.

Gillet, J.D., 1974. Direct and indirect influences of temperature on the transmission of parasites from insects to man. p79-95 in: *The Effects of Meteorological Factors upon Parasites.* (A.E.R. Taylor and R. Muller, eds.) Oxford: Blackwell Scientific Publications.

Giorgi, F., and L.O. Mearns. 1991. Approaches to the simulation of regional climate change: A review. *Rev. of Geophysics* 29:191-216.

Glacken, C.J. 1967. *Traces on the Rhodian Shore; Nature and Culture in Western Thought from Ancient Times to the End of the Eighteenth Century.* Berkeley: University of California Press.

Glantz, M.H. 1996. *Currents of Change: El Niño's Impact on Climate and Society.* Cambridge: Cambridge University Press.

Glass G.E., J.E. Cheek, J.A. Patz, T.M. Shields, T.J. Doyle, D.A. Thoroughman, D.K. Hunt, R.E. Enscore, K.L. Gage, C. Irland, C.J. Peters, and R. Bryan. 2000. Using remotely sensed data to identify areas at risk for hantavirus pulmonary syndrome. *Emerging Infectious Diseases* 6:238-247.

Goddard, L, S.J. Mason, S.E. Zebiak, C.F. Ropelewski, R. Basher, and M.A. Cane. 2001. Current approaches to seasonal to interannual climate predictions. *International Jour. of Climatology* (In Press).

Gordon, S.M., C.J. Carlyn, L.J. Doyle, C.C. Knapp, D.L. Longworth, G.S. Hall, and J.A. Washington. 1996. The emergence of Neisseria gonorrhoeae with decreased susceptibility to ciprofoxacin in Cleveland, Ohio: Epidemiology and risk factors. *Annals of Internal Medicine* 125:465-470.

Grabherr, G., M. Gottfried, and H. Pauli. 1994. Climate effects on mountain plants. *Nature* 369:448-449.

Grant, E. 1994. *Planets, Stars and Orbs: The Medieval Cosmos, 1200 -1687.* Cambridge: Cambridge University Press.

Gray, W.M., C.W. Landsea, P.W. Mielke, Jr., and K.J. Berry. 1999. Extended range forecast of Atlantic seasonal hurricane activity and U.S. landfalling strike probability for 2000. Dept. of Atmos. Sci., Colo. State Univ., Ft. Collins, Colo., 8 December, 19 pp.

Greenwood, B.M., I.S. Blakebrough, A.K. Bradley, S. Wali, and H.C. Whittle. 1984. Meningococcal disease and season in sub-Saharan Africa. *Lancet* 1:1339-1342.

Gubler, D.J. 1988. Dengue. Pp. 223-260. in *The Arboviruses: Epidemiology and Ecology, Vol. II,* T.P. Monath, ed. Boca Raton, Fla.: CRC Press.

Guillet P., M.C. Germain, T. Giacomini, F. Chandre, M. Akogbeto, O. Faye, A. Kone, L. Manga, and J. Mouchet. 1998. Origin and prevention of airport malaria in France. *Tropical Medicine and International Health* 3:700-705.

Haas, C.N, J.B. Rose, and C.P. Gerba, eds. 1999. *Quantitative Microbial Risk Assessment.* New York: John Wiley & Sons.

Hairston, N.G., 1973. The dynamics of transmission, pp250-336 in: N. Ansari, ed. *Epidemiology and control of schistosomiasis.* Basel: Karger.

Halstead, S.B. 1988. Pathogenesis of dengue: Challenges to molecular biology. *Science* 239:476-481.

Hammer, G., D.P. Holzworth, and R. Stone. 1996. The value of skill in seasonal climate forecasting to wheat crop management in a region with high climatic variability. *Aust. Journ. of Agric. Res.* 47:717-737.

Hammer, G. 2000. Applying seasonal climate forecasts in agricultural and natural ecosystems—A synthesis. Chapter 27 in *Applications of Seasonal Climate Forecasting to Agricultural and Natural Ecosystems—the Australian Experience.* G. Hammer, N. Nicholls, and C. Mitchell, eds. The Netherlands: Kluwer.

Hancock, C.M., J.B. Rose, and M. Callahan. 1998. Crypto and Giardia in U.S. groundwater. *J. Am. Water Works Assoc.* 90(3):58-61.

Hannaway, C.C. 1972. The *Société Royale de Médecine* and epidemics in the *Ancien Régime. Bulletin of the History of Medicine* 46:257-273.

Hannaway, C., 1993. Environment and miasmata. Pp. 292-308 in W.F. Bynum and R. Porter, eds. *Companion Encyclopedia of the History of Medicine.* Ithaca, N.Y.: Cornell University Press.

Harte, J., and R. Shaw. 1995. Shifting dominance within a montane vegetation community: Results of a climate warming experiment. *Science* 267:876-880.

Hay, S.I., S.E. Randolph, and D.J. Rogers. 2000. *Remote Sensing and Geographical Information Systems in Epidemiology.* New York: Academic Press.

Hirsch, A. 1883-1886. *Handbook of Geographical and Historical Pathology,* translated by Charles Creighton. The volumes. London.

Hornick, R.B, S.I. Music, R. Wenzel, R. Cash, J.P. Libonati, and T.E. Woodward. 1971. The Broad Street pump revisited: Response of volunteers to ingested cholera vibrios. *Bull. N.Y. Acad. Med.,* 47(10):1181-1191.

IPCC (Intergovernmental Panel on Climate Change). 2001. Climate Change 2001: The Scientific Basis. WG I contribution to the IPCC Third Assessment Report. Summary for Policymakers. Available online at: *http://www.ipcc.ch/* (Full technical report to be released in 2001.)

Janssen, M., and P. Martens. 1997. Modelling malaria as a complex adaptive system. *Artificial Life* 3:213-236.

Jetten, T.H., and D.A. Focks. 1997. Changes in the distribution of dengue transmission under climate warming scenarios. *Am. J. Trop. Med. Hyg.* 57:285-297.

Jia, X.Y., T. Briese, I. Jordan, A. Rambaut, H.C. Chi, J.S. Mackenzie, R.A. Hall, J. Scherret, W.I. Lipkin. 1999. Genetic analysis of West Nile New York 1999 encephalitis virus. *Lancet* 354:1971-1972.

Jordanova, L.J. 1979. Earth science and environmental medicine: The synthesis of the late enlightenment. Pp. 119-146 in *Images of the Earth: Essays in the History of the Environmental Sciences.* L.J. Jordanaova and R. Porter, eds. London: British Society for the History of Science.

Justice, C.O., J.R.G. Townshend, B.N. Holben, and C.J. Tucker. 1985. Analysis of the phenology of global vegetation using meteorological satellite data. *Int. J. Remote Sens.* 8:1271-1318.

Kaneko, T., and R.R. Colwell. 1973. Ecology of *Vibrio parahaemolyticus* in Chesapeake Bay. *J. Bacteriol.* 113:24-32.

Kattenberg, A., F. Giorgi, H. Grassl, G.A. Meehl, J.F.B. Mitchell, R.J. Stouffer, T. Tokioka, A.J. Weaver, and T.M.L. Wigley. Climate models—projections of future climate. Pp. 285-357. in *Climate Change 1995: The Science of Climate Change*, Contribution of Working Group I to the Second Assessment Report of the Intergovernmental Panel on Climate Change. Cambridge: Cambridge University Press.

Kenyon, T.A., S.E. Volway, W.W. Ihle, I.M. Onorato, and K.G. Castro. 1996. Transmission of multi-drug resistant Mycobacterium tuberculosis during a long airplane flight. *New England Journal of Medicine* 334:933-938.

Kinsella, R.A. 1935. Report on the St. Louis Outbreak of Encephalitis, U.S. Treasury Department, Public Health Bulletin, No. 214, 117 pp.

Knapp, J.D., K.K. Fox, D.L. Trees, and W.L. Whittington. 1997. Fluoroquinoline resistance in Neisseria gonorrhoeae. *Emerging Infectious Diseases* 3:33-39.

Krause, R.M. 2001. Microbes and emerging infections: The compulsion to become something new. *Am. Soc. Microbiol. News* 67(1):15-20.

Krieger, N. 1994. Epidemiology and the web of causation: Has anybody seen the spider? *Social Science and Medicine* 39:881-903.

Kunitz, S.J. 1987. Explanations and ideologies of mortality patterns. *Population and Development Review* 13:379-407.

Kutzbach, G. 1979. *The Thermal Theory of Cyclones: A History of Meteorological Thought in the Nineteenth Century.* Historical Monograph Series. Boston: American Meteorological Society.

Lanciotti, R.S., J.T. Rochrig, V. Deubel, J. Smith, M. Parker, K. Steele, B. Crise, K.E. Volpe, M.B. Crabtree, J.H. Scherret, R.A. Hall, J.S. MacKenzie, C.B. Cropp, B. Panigrahy, E. Ostlund, B. Schmitt, M. Malkinson, C. Banet, J. Weissman, N. Komar, H.M. Savage, W. Stone, T. McNamara, and D.J. Gubler. 1999. Origin of the West Nile virus responsible for an outbreak of encephalitis in the northeastern United States. *Science* 286:2333-2337.

Langford, I.H., and G. Bentham. 1995. The potential effects of climate change on winter mortality in England and Wales. *Int. J. Biometeorol.* 38(3):141-147.

Levins, R., T. Auerbach, U. Brinckmann, I. Eckhardt, P. Epstein, C. Makhoul, C. Albuquerque de Passos, C. Puccia, A. Spielman, and M. Wilson. 1994. The emergence of new diseases. *American Scientist* 82:52-60.

Lindblade, K.A., E.D. Walker, A.W. Onapa, J. Katungu, and M.L. Wilson. 1999. Highland malaria in Uganda: Prospective analysis of an epidemic associated with El Nino. *Trans. Roy. Soc. Trop. Med. Hyg.* 39:480-487.

Lindsay, S.W., L. Parson, and C. J. Thomas. 1998. Mapping the ranges and relative abundance of the two principal African malaria vectors, *Anopheles gambiae sensu stricto* and *An. arabiensis*, using climate data. *Proceedings of the Royal Society of London* 265:847-854.

Linthicum, K.J., A. Anyamba, C.J. Tucker, P.W. Kelley, M.F. Myers, and C.J. Peters. 1999. Climate and satellite indicators to forecast Rift Valley fever epidemics in Kenya. *Science* 285:397-400.

Linthicum K.J., C.L. Bailey, F.G. Davies, A. Kairo, and T.M. Logan. 1988. The horizontal distribution of Aedes pupae and their subsequent adults within a flooded dambo in Kenya: implications for Rift Valley fever virus control. *J. Am. Mosq. Control Assoc.* 4:551-554.

Lisle, J.T., and J.B. Rose. 1995. *Cryptosporidium* contamination of water in the USA and UK: A mini-review. *J. Water SRT—Aqua* 44:103-117.

Lobitz, B., L. Beck, A. Huq, B. Wood, G. Fuchs, A.S.G. Faruque, and R. Colwell. 2000. Climate and infectious disease: Use of remote sensing for detection of *Vibrio cholerae* by indirect measurement *Proc. Natl. Acad. Sci.* 97:1438-1443.

Lorenz, E.N. 1982. Atmospheric predictability experiments with a large numerical model. *Tellus* 34:505-513.

Lorenz, E.N. 1993. *The Essence of Chaos.* Seattle: University of Washington Press.

MacKenzie, W.R., N.J. Hoxie, M.E. Proctor, S. Gradus, K.A. Blair, D.E. Peterson, J.J. Kazmierczak, D.G. Addiss, K.R. Fox, J.B.Rose, and J.P. David. 1994. A massive outbreak in Milwaukee of *Cryptosporidium* infection transmitted through the public water supply. *N. Engl. J. Med..* 331(3):161-167.

Martens, P. 1998. Health and Climate Change: Modelling the impacts of global warming and ozone depletion. London: Earthscan.

Martens P., R.S. Kovats, S. Nijhof, P. deVries, M.T. J. Livermore, D.J. Bradley, J. Cox, and A.J. McMichael. 1999. Climate change and future populations at risk of malaria. *Global Environmental Change* 9:89-107.

Mason, S.J., L. Goddard, N.E. Graham, E. Yulaeva, L. Sun, and P.A. Arkin. 1999. The IRI seasonal climate prediction system and the 1997/98 El Niño event. *Bull. Amer. Meteor. Soc.* 80:1853-1873.

Mata L. 1994. Cholera El Tor in Latin America, 1991-1993. *Annals of the New York Academy of Science* 740:55-68.

May, J.M. 1958. *The Ecology of Human Disease.* New York: M.D. Publications.

Mayer, J.D. 2000. Geography, ecology, and emerging diseases. *Social Science and Medicine* 50:937-952.

McCutchan, M.H., B. Meisner, F. Fujioka, J. Benoit, and B. Ly. 1991. Monthly fire weather forecasts. *Fire Management Notes.* 52(3):41-47.

McIntosh, B.M., and P.G. Jupp. 1981. Epidemiological aspects of Rift Valley fever in South Africa with references to vectors. *Contrib. Epidemiol. Biostat.* 3:92-99.

McMichael, A.J. 1994. Molecular epidemiology: new pathway or new travelling companion? *American Journal of Epidemiology* 140: 1-11.

Meegan, J.M. and C.L. Bailey. 1988. Rift Valley fever. Pp. 51-76 in *Arboviruses Epidemiology and Ecology,* Vol IV, T. P. Monath, ed. Boca Raton, FL: CRC Press.

Meinke, H., and Z. Hochman. 2000. Using seasonal climate forecasts to manage dryland crops in northern Australia—experience from the 1997/98 season. in *Applications of Seasonal Climate Forecasting to Agricultural and Natural Ecosystems*—the Australian Experience, G. Hammer, N. Nicholls, and C. Mitchell, eds. The Netherlands: Kluwer.

Miller B.R., R.S. Nasci, M.S. Godsey, H.M. Savage, J.J. Lutwama, R.S. Lanciotti, and C.J. Peters. 2000a. First field evidence for natural vertical transmission of West Nile virus in Culex univittatus complex mosquitoes from Rift Valley province, Kenya. *Am. J. of Tropical Medicine and Hygiene* 62(2):240-246.

Miller J.M., T.W. Tam, S. Maloncy, K. Fukuda, N. Cox, J. Hockin, D. Kertesz, A. Klimov, and M. Cetron. 2000b. Cruise ships: high-risk passengers and the global spread of new influenza viruses. *Clin Infect Dis.* 31:433-438.

Mjelde, J.W., T.N. Thompson, and C.G. Coffman. 1997. Using Southern Oscillation information for determining corn and sorghum profit-maximizing input levels in east-central Texas. *J. of Production Agriculture.* 10:168-178.

Mobarak, A.B. 1982. The schistosomiasis problem in Egypt. *Am. J. of Tropical Medicine and Hygiene* 31:87-91.

Mollison, D. ed. 1995. *Epidemic Models: Their Structure and Relation to Data.* Cambridge: Cambridge University Press.

MMWR, 2000. Update: West Nile Virus activity—Eastern United States. *Morb. Mortal. Wkly. Rep.* 49:1044-1047.

Molineaux, L. 1988. The epidemiology of human malaria as an explanation of its distribution, including some implications for its control. In: W.H. Wernsdorfer, I. McGregor, eds. *Malaria: Principle and Practice of Malariology.* Edinburgh: Chruchill Livingstone

Monath, T.P. 1980. Chapter 6, Epidemiology, pp. 263-266 in *St. Louis Encephalitis*, T.P. Monath (ed.) Washington, DC. American Public Health Association.

Monath T.P. and T.F. Tsai. 1987. St. Louis encephalitis: Lessons from the last decade. *Am. J. Trop. Med. Hyg.* 37(3):40S-59S.

Monmonier, M. 1999. *Air Apparent: How Meteorologists Learned to Map, Predict, and Dramatize Weather.* Chicago: University of Chicago Press.

Moore, C.G., and C.J. Mitchell. 1997. Aedes albopictus in the United States: Ten Year Presence and public health implications. *Emerging Infectious Diseases* 3:329-334.

Moore C.G., B.L. Cline, E. Ruiz-Tibén, D. Lee, H. Romney-Joseph, and E. Rivera-Correa. 1978. *Aedes aegypti* in Puerto Rico: Environmental determinants of larval abundance and relation to dengue virus transmission. *Am. J. Trop. Med. Hyg.* 27:1225-1231.

Morse, S.S. 1995. Factors in the emergence of infectious diseases. *Emerging Infectious Diseases* 1(1).

Motes M.L., A. DePaola, D.W. Cook, J.E. Veazey, J.C. Hunsucker, W.E. Garthright, R.J. Blodgett, and J. Chirtel. 1998. Influence of water temperature and salinity on Vibrio vulnificus in Northern Gulf and Atlantic Coast oysters (*Crassostrea virginica*). *Applied and Environmental Microbiology* 64(4):1459-1465.

Mount, G.A., D.G. Haile, and E. Daniels. 1997. Simulation of management strategies for the black-legged tick (Acari: Ixodidae) and the Lyme disease spirochete, Borrelia burgdorferi. *J. Med. Entomol.* 90:672-683.

Musil, 1999. Physicians for Social Responsibility (Robert Musil, Exec. Director). Press Conference, National Press Club, Washington, D.C., October 9, 1999.

Myeni, R., C. Keeling, C. Tucker, G. Asrar, and R. Nemani. 1997. Increased plant growth in the northern high latitudes from 1981 to 1991. *Nature* 386:698-699.

Nasci, R.S. and C.G. Moore. 1998. Vector-borne disease surveillance and natural disasters. *Emerging Infectious Diseases* 4:333-334.

National Research Council. 1983. *Risk Assessment in the Federal Government: Managing the Process.* National Academy Press, Washington, D.C.

National Research Council. 1992. *Emerging Infections: Microbial Threats to Health in the U.S.* National Academy Press, Washington, D.C.

National Research Council. 1998. *Making Climate Forecasts Matter.* P.C. Stern and W.E. Easterling, eds. National Academy Press, Washington, D.C.

National Research Council. 2000. *Reconciling Observations of Global Temperature Change.* National Academy Press, Washington, D.C.

Nebeker, F. 1995. *Calculating the Weather: Meteorology in the 20th Century.* New York: Academic Press.

Nichol, S.T, C.F. Spiropoulou, S. Morzunov, P.E. Rollin, T.G. Ksiazek, H. Feldmann, A. Sanchez, J. Childs, S. Zaki, and C.J.Peters. 1993. Genetic identification of a hantavirus associated with an outbreak of acute respiratory illness. *Science* 262:615-618.

Noji, E.K., ed. 1997. *The Public Health Consequences of Disasters*. New York: Oxford University Press.

Nutton, V. 1983. The seeds of disease: An explanation of contagion and infection from the Greeks to the Renaissance. *Medical History* 27:1-34.

O'Neill, K.R., S.H. Jones, and D.J. Grimes. 1992. Seasonal incidence of *Vibrio vulnificus* in the Great Bay Estuary of New Hampshire and Maine. *Appl. Environ. Microbiol.* 58:3257-3262.

Pan-American Health Organization (PAHO). 1998. *El Niño and its Impact on Health*. Report presented to the 122nd Executive Assembly. Document CE122/10.

Parmenter R.R., J.W. Brunt, D.I. Moore, and S. Ernest. 1993. The hantavirus epidemic in the Southwest: Rodent population dynamics and the implications for transmission of hantavirus-associated adult respiratory distress syndrome (HARDS) in the Four Corners region. Sevilleta LTER Publication No. 41.

Parmesan, C. 1996. Climate and species' range. *Nature* 382:765.

Parson, E.A., and K. Fisher-Vanden. 1997. Integrated assessment of global climate change. *Annual Review of Energy and the Environment* 22.

Pascual, M., X. Rodo, S.P. Ellner, R. Colwell, and M.J. Bouma. 2000. Cholera dynamics and El-Niño-Southern Oscillation. *Science* 289:1766-1769.

Patz, J.A., W.J.M. Martens, D.A. Focks, and T.H. Jetten. 1998. Dengue fever epidemic potential as projected by general circulation models of global climate change. *Environ. Health Perspectives* 106:147-152.

Pelling, M. 1993. Contagion/germ theory/specificity. in *Companion Encyclopedia of the History of Medicine*, Two volumes. W. F. Bynum and R. Porter, eds., London: Routledge.

Peters, C.J. and J.M. Meegan. 1981. Rift Valley fever. Pp. 403-420m in *CRC Handbook Series in Zoonoses*. J. H. Steele, ed. Boca Raton, Fla.: CRC Press.

Pickering, H., M. Okongo, A. Ojwiya, D. Yirrell and J. Whitworth. 1997. Sexual networks in Uganda: Casual and commercial sex in a trading town. *AIDS Care* 9:199-207.

Pielke, R.A., Jr., and R.A. Pielke, Sr. 1997. *Hurricanes: Their Nature and Impacts on Society*. Chichester: John Wiley & Sons.

Pielke, R.A., Sr., R.L. Walko, L.T. Steyaert, P.A. Vidal, C.T. Liston, W.A. Lyons, and T.N. Chase. 1999. The influence of anthropogenic landscape changes on weather in south Florida. *Monthly Weather Review* 127:1663-1674.

Porter, T.M. 1986. *The Rise of Statistical Thinking: 1820-1900*. Princeton, N.J.: Princeton University Press.

Price, M., and N. Waser. 1998. Effects of experimental warming on plant reproductive phenology in a subalpine meadow. *Ecology* 79:1261-1271.

Prothero, R.M. 1965. *Migrants and Malaria*. London: Longmans.

Pullan, B. 1992. Plague and perceptions of the poor in early modern Italy. Pp. 101-123 in *Epidemics and Ideas: Essays on the Historical Perception of Pestilence*. T. Ranger and P. Slack, eds. Cambridge: Cambridge University Press.

Pulwarty, R.S., and W.E. Riebsame. 1997. The political ecology of vulnerability to hurricane-related hazards. Pp 185-214 in *Hurricanes: Climate and Socio-economic Impacts*. H. Diaz, and R. S. Pulwarty, eds. Heidelberg: Pringer-Verlag.

Pyle, G.F. 1969. The diffusion of cholera in the United States in the nineteenth century. *Geographical Analysis,* 1:59-75.

Quinn, W.H., and V.T. Neal. 1992. The historical record of El Niño events. Pp. 623-648 in *Climate Since A.D. 1500*. R.S. Bradley and P.D. Jones, eds. London and New York: Routledge.

Reeves, W.C., J.L. Hardy, W.K. Reisen, and M.M. Milby. 1994. Potential effect of global warming on mosquito-borne arboviruses. *Journal of Medical Entomology* 31:323–332.

Regli, S., J.B. Rose, C.N. Haas, and C.P. Gerba. 1991. Modeling the risk from Giardia and viruses in drinking water. *J. Am. Water Works Assoc.* 83:76-84.

Reisen W.K., M.M. Milby, S.B. Presser, and J.L. Hardy. 1992. Ecology of mosquitoes and St. Louis encephalitis virus in the Los Angeles Basin of California, 1987-1990. *J. Med. Entomol.* 29:582-598.

Reisen W.K, J.O. Lundstrom, T.W. Scott, B.F. Eldridge, R.E. Chiles, R. Cusack, V.M. Martinez, H.D. Lothrop, D. Gutierrez, S.E. Wright, K. Boyce, and B.R. Hill. 2000. Patterns of avian seroprevalence to western equine encephalomyelitis and Saint Louis encephalitis viruses in California, USA. *J. Med. Entomol.* 37:507-527.

Reiter, P. 2000. From Shakespeare to Defore: Malaria in England in the Little Ice Age. *Emerging Infectious Diseases* 6(1):1-11.

Riley, J.C. 1987. *The Eighteenth-Century Campaign to Avoid Disease.* New York: St. Martin's Press.

Roads, J.O., S.C. Chen, F.M. Fujioka, and R.E. Burgan. 2000. Development of a seasonal fire severity forecast for the contiguous United States. Third Symposium on Fire and Forest Meteorology. American Meteorological Society, 10-12 January 2000, Long Beach, Calif.

Robard, D. 1981. The role of regional body temperature in the pathogenesis of disease. *New England Journal of Medicine* 305:808-814.

Rogers, D.J., and S.E. Randolph. 1993. Distribution of tsetse and ticks in Africa: Past, present and future. *Parasitology Today* 9:266-271.

Rogers, D.J., and S.E. Randolph. 2000. The global spread of malaria in a future, warmer world. *Science* 289:1763-1766.

Ropelewski C.F., and M.S. Halpert. 1987. Global and regional scale precipitation patterns associated with the El Niño/Southen Oscillation. *Mon. Wea. Rev.* 115:1606-1626.

Rose, J.B. 1997. Environmental ecology of *Cryptosporidium* and public health implications. *Annual Review of Public Health* 4.

Rose, J.B. and C.P. Gerba. 1991. Use of risk assessment for development of microbial standards. *Water Sci. Technol.* 24:29-34.

Rose, J., C.N. Haas, and S. Regli. 1991. Risk assessment and control of waterborne giardiasis. *Am. J. Pub. Health* 1:709-713.

Rose, J.B., S. Daeschner, D.R. Easterling, F.C. Curriero, S. Lele, and J.A. Patz. 2000. Climate and waterborne disease outbreaks. *J. Am. Water Works Assoc.* 9:77-87.

Rosenfeld, D. 2000. Suppression of rain and snow by urban and industrial air pollution. *Science.* 287: 1793-1796.

Rotmans, J. 1998. Methods for IA: the challenges and opportunities ahead. *Environmental Modeling and Assessment* 3(3):155-179.

Rotmans, J. and M. Van Asselt. 1999. Integrated assessment modelling. Climate Change: An Integrated Perspective. in P. Martens and J. Rotmans, eds. Dordrecht: Kluwer Academic Publishers.

Saleska, S., J. Harte, and M. Torn. 1999. The effect of experimental ecosystem warming on net CO_2 fluxes in a montane meadow. *Global Change Biology* 5(2):125-142.

Saunders, M.A., R.E. Chandler, C.J. Merchant, and F.P. Roberts. 2000. Atlantic hurricanes and NW Pacific typhoons: ENSO spatial impacts on occurrence and landfall. *Geophys. Res. Lett.* 27:1147-1150.

Schneider, S. 1997. Integrated assessment modeling of climate change: Transparent rational tool for policy making or opaque screen hiding value-laden assumptions? *Environmental Modelling and Assessment* 2(4):229-250.

Schneider, E., R.A. Hajjeh, R.A. Spiegel, R.W. Jibson, E.L. Harp, G.A. Marshall, R.A. Gunn, M.M. McNeil, R.W. Pinner, R.C. Baron, L.C. Hutwagner, C. Crump, L. Kaufman, S.E. Reef, G.M. Feldman, D. Pappagianis, and S.B. Werner. 1997. A coccidioidomycosis outbreak following the Northridge, California, earthquake. *J. Am. Med. Assoc.* 277: 904-908.

Schulman, J.L., and E.D. Kilbourne. 1963. Experimental transmission of influenza virus infection in mice: Some factors affecting the incidence of transmitted infection. *J. Exp. Med.* 118:267-275.

Seargeant, F. 1982. *Hippocratic Heritage: A History of Ideas about Weather and Human Health.* New York: Pergamon Press.

Sethi, I., and A.K. Jain, ed. 1991. *Artificial Neural Networks and Pattern Recognition: Old and New Connections.* Amserdam: Elsevier.

Shen, K., and J. Harte. 2000. Ecosystem Climate Manipulations. Pp. 353-369 in *Methods in Ecosystem Science.* O. Sala, R. Jackson, H. Mooney, and R. Howarth eds. New York: Springer-Verlag.

Sheppard, P.M., W.W. Macdonald, R.J. Tonn, and B. Grabs. 1969. The dynamics of an adult population of *Aedes aegypti* in relation to dengue haemorrhagic fever in Bangkok. *J. Anim. Ecol.* 38:661-702.

Shimshony, A., and R. Barzilai. 1983. Rift Valley fever. *Adv. Vet. Sci. Comp. Med.* 27:347-425.

Shope R, 1991. Global climate change and infectious diseases. *Environmental Health Perspectives* 96:171-174.

Shope, R.E., and T.F. Tsai. 1998. Diseases transmitted primarily by arthropod vectors: Viral infections, p. 290, in Wallace RB (ed.) *Public Health and Preventive Medicine*, 14th ed. Stamford, Conn.: Appleton & Lange.

Slack, P. 1985. *The Impact of Plague in Tudor and Stuart England.* London: Routledge and Kegan Paul.

Smith, H.V., and J.B. Rose. 1998. Waterborne cryptosporidiosis: Current status. *Parasitology Today* 14(1):14-22.

Smith, W.D., 1979. *The Hippocratic Tradition.* Ithaca, N.Y.: Cornell University Press.

Smith C.E., R.R. Beard, H.G. Rosenberger, and E.G. Whiting. 1946. Effect of season and dust control on coccidioidomycosis. *J. Am. Med. Assoc.* 132:833-838.

Solow, A.R., R.F. Adams, K.J. Bryant, D.M. Legler, J.J. O'Brien, B.A. McCarl, W. Nayda, and R. Weiher. 1998. The value of improved ENSO prediction to U.S. Agriculture. *Climatic Change* 39:47-60.

Southwood, T.R.E., G. Murdie, M. Yasuno, R.J. Tonn, and P.M. Reader. 1972. Studies on the life budget of *Aedes aegypti* in Wat Samphaya, Bangkok, Thailand. *Bull. World Health Organization* 46:211-226.

Stone, R.C., G.C. Hammer, and T. Marcussen. 1996. Prediction of global rainfall probabilities using phases of the Southern Oscillation Index. *Nature* 384:252-255.

Susser, M. 1985. Epidemiology in the United States after World War II: the evolution of technique. *Epidemiologic Reviews* 7:47-77.

Susser, M. 1986. The logic of Sir Karl Popper and the practice of epidemiology. *American Journal of Epidemiology* 124:711-718.

Susser, M., and E. Susser. 1996. Choosing a future for epidemiology, I: Eras and paradigms. *American Journal of Public Health* 86:668-673.

Susser, M. 1998. Does risk factor epidemiology put epidemiology at risk? Peering into the future. *Journal of Epidemiology and Community Health* 52:608-618.

Sutherst, R.W., G.F. Maywald, and D.B. Skarratt. 1995. Predicting insect distributions in a changed climate. Pp.60-91 in *Insects: A Changing Environment*, R. Harrington and N.E. Stork, eds. 17th Symposium of the Royal Entomological Society of London.

Thomas, C. and J. Lennon. 1999. Birds extend their range northwards. *Nature* 399: 213.

Tompkins, A.M. 1986. Protein-energy malnutrition and risk of infection. *Proceedings of the Nutrition Society* 45:289-304.

Toon, B. 2000. Perspectives: Atmospheric science: How pollution suppresses rain. *Science* 287:1763-1765.

Tucker, C., H. Dregne, and W. Newcomb. 1991. Expansion and contraction of the Sahara Desert from 1980 to 1990. *Science* 253:299-301.

Tulu, A. 1996. Determinants of malaria transmission in the highlands of Ethiopia: The impact of global warming on morbidity and mortality ascribed to malaria. Thesis, London School of Hygiene and Tropical Medicine, University of London.

U. S. Environmental Protection Agency. 1989. National primary drinking water regulations; filtration and disinfection; turbidity; Giardia lamblia, viruses, Legionella, and heterotrophic bacteria, *Federal Register* 54(124):27486-27541.

Vandenbroucke, J.P. 1989. Those who were wrong. *American Journal of Epidemiology* 130:3-5.

Walford, C. 1879. Famines of the World, Past and Present. *Journal of the Royal Statistical Society* 42:79-275.

Walsh, J.F., D.H. Molineaux, and M.H. Birley. 1993. Deforestation: effects on vector-borne disease. *Parasitology* 106:55-75.

Watkins, W.D., and V.J. Cabelli. 1985. Effect of fecal pollution on *Vibrio parahaemolyticus* densities in an estuarine environment. *Appl. Environ. Microbiol.* 49:1307-1313.

Watson, G., R.E. Shope, and M. Kaiser. 1972. An ectoparasite and virus survey of migratory birds in the eastern Mediterranean. *Academy of Sciences of the USSR, Siberian Branch*, pp.176-180.

Webb, T., III., 1986. Is vegetation in equilibrium with climate? How to interpret late-Quaternary pollen data. *Vegetation* 67:75-91.

Webster R.G., W.J. Bean, O.T. Gorman, T.M. Chambers, and Y. Kawaoka. 1992. Evolution and ecology of influenza A viruses. *Microbiol Rev.* 56(1):152-179.

Weiher, R.F., ed. 1999. *Improving El Nino Forecasting: The Potential Economic Benefits.* U.S. Department of Commerce: Washington, D.C.

Weniger, B.G., M.J. Blaser, J. Gedrose, E.C. Lippy, and D.D. Juranek. 1983. An outbreak of waterborne giardiasis associated with heavy water runoff due to warm weather and volcanic ashfall. *Amer. J. Public Hlth.* 73:868-872.

White, K.L. 1991. *Healing the Schism: Epidemiology, Medicine, and the Public's Health.* New York: Springer.

Whitnah, D.R. 1965. *A History of the United States Weather Bureau.* Urbana: University of Illinois Press.

WHO, 1985. Ten years of oncho. control in West Africa: review of work of the OCP in the Volta River Basin area from 1974 to 1984. OCP/GVA/85.1B, WHO, Geneva Switzerland, 113 pp.

WHO, 1998a Dengue and Dengue Haemorragic Fever. Fact Sheet No 117. Available online at: *http://www.who.int/inf-fs/en/fact117.html*

WHO, 1998b. Malaria. Fact Sheet No 94. Available online at: *http://www.who.int/inf-fs/en/fact094.html*

WHO, 1999a . Influenza. Fact Sheet No 211. Available online at: *http://www.who.int/inf-fs/en/fact211.html*

WHO, 1999b. The WHO Influenza Programme. Available online at: *http://www.who.int/inf-fs/en/fact212.html*

WHO, 1999c. Weekly Epidemiological Record. 6 August. 74: 257-264. Available online at *http://www.who.int/wer/ http://www.who.int/wer/*

Wigley, T.M.L., M.J. Ingram, and G. Farmer, eds. 1992. *Climate and History.* Cambridge: Cambridge University Press.

Wilson, M.L. 1994. Rift Valley fever virus ecology and the epidemiology of disease emergence. *Ann. N. Y. Acad. Sci.* 740:169-180.

Wilson M.L. 1998. Distribution and abundance of Ixodes scapularis (Acari:Ixodidae) in North America: ecological processes and spatial analysis. *J. Med. Entomol.* 35:446-457.

Winslow, C. A-E. 1980. *The Conquest of Epidemic Diseases.* Madison: University of Wisconsin Press.

Wilson, M., R. Levins, and A. Spielman. 1994. *Disease in Evolution: Global Changes and Emergence of Infectious Diseases.* New York: New York Academy of Sciences. 503 pp.

Woodruff, D. 1992. Wheatman, a decision support system for wheat management in subtropical Australia. *Aust. J. Agric. Res.* 43:1485-1499.

World Meteorological Organization. 1999. *The 1997-1998 El Nino Event: A Scientific and Technical Retrospective.* WMO: Geneva.

Wuhib, T., T.M. Silva, R.D. Newman, L.S. Garcia, M.L. Pereira, C.S. Chaves, S.P. Wahlquist, R.T. Bryan, and R.L. Guerrant. 1994. Cryptosporidial and microsporidial infections in human immunodeficiency virus-infected patients in northeastern Brazil. *Journal of Infectious Diseases* 170:494-497.

A

Biographical Sketches of Committee Members

Donald Burke *(Chair)* is a professor in the Department of International Health and director of the Center for Immunization Research at the Johns Hopkins University School of Public Health. Previously he served as chief of the Department of Virus Diseases and as director of the Division of Retrovirology at the Walter Reed Army Institute of Medical Research. He also served as director of the U.S. Military HIV/AIDS Research Program and as chief of the Department of Virology for the Armed Forces Research Institute of Medical Sciences in Bangkok, Thailand. His past research focused on tropical viral diseases such as dengue and encephalitis; and his current major interests are molecular epidemiology and the evolution of human viruses. He served on the National Research Council's Roundtable for the Development of Drugs and Vaccines Against AIDS and is past president of the American Society of Tropical Medicine.

Ann Carmichael has an M.D. and a Ph.D. from Duke University and is currently an associate professor in the Department of History at Indiana University. She teaches classes in the history of epidemics and human infections and has published numerous articles and book chapters on these topics. She has served on the editorial boards of the *Bulletin of the History of Medicine*, the *Journal of the History of Medicine and Allied Sciences*, the *American Historical Review*, and the *Cambridge World History of Human Disease*.

Dana Focks is a senior research scientist with the U.S. Department of Agriculture's Center for Medical, Agricultural, and Veterinary Entomology. His early work involved biocontrol and the development of computer simulation models to help control *Ae. aegypti*, the vector of dengue hemorrhagic fever, and vectors of Venezuelan equine encephalitis. Later work focused on the development of weather-driven epidemiological models of dengue transmission. These models are currently used in numerous countries around the world including the U.S. Department of Defense, the World Health Organization and the Pan American Health Organization. Current efforts have largely been directed toward developing validated assessments of the potential consequences of climate change and El Niño Southern/Oscillation events for dengue and lyme disease.

Darrell Jay Grimes is director of the Institute of Marine Sciences and professor of coastal sciences at the University of Southern Mississippi. His research interests include microbiological quality of water resources and bacterial genetics in natural environments. Previously, he served as director of the University of New Hampshire's Institute of Marine Science and Ocean Engineering and its Sea Grant College Program and as a microbiologist for U.S. Department of Energy's Environmental Sciences Division. He has served on many different advisory bodies for federal agencies and universities, as well as the National Research Council's Ocean Studies Board.

John Harte is a professor in the Energy and Resources Group and the Department of Environmental Science, Policy, and Management at the University of California, Berkeley. His research interests include climate-ecosystem feedback processes, theoretical ecology with an emphasis on elucidating patterns of sealing in the distribution and abundance of species, causes and consequences of declining biodiversity, biogeochemical processes and their disruption, and the role of ecological integrity in human society. Dr. Harte is an associate editor of the *Annual Review of Energy and the Environment*. He served on the National Research Council's Committee on Scientific Issues in the Endangered Species Act and previously served on four other NRC committees.

Subhash Lele is a professor in the Department of Mathematical Sciences at the University of Alberta. Previously he was with the Department of Biostatistics at the Johns Hopkins University School of Public Health, where he served as a director of the university's Program on Health Effects of Global Environmental Change. Dr. Lele holds a Ph.D. in statistics, and his research interests include spatial statistics and geographic information systems with applications in ecology, environmental sciences, and public health, as well as more theoretical work in the foundations of statistics. He recently helped develop models of vector-borne disease patterns and the impacts of climate change.

Pim Martens holds degrees in biological and environmental health sciences, and has a Ph.D. in mathematics from Maastricht University, Netherlands. He worked on the global dynamics and sustainable development project launched in

1992 by the Dutch National Institute of Public Health and the Environment. Since 1998 he has served as a senior researcher at the university's International Centre for Integrative Studies, where he directs the Global Assessment Centre. He is also editor in chief of the international journal Global Change and Human Health and editor of a book series on resurgent and emerging infectious diseases. Dr. Martens is project leader of various national and international projects on global change and human health; he has contributed assessment reports to the Intergovernmental Panel on Climate Change, the World Health Organization and the United Nations Environment Programme, and is a member on several scientific advisory committees.

Jonathan Mayer is a professor of geography; adjunct professor of medicine, Division of Infectious Diseases; adjunct professor of family medicine; and adjunct professor in the School of Public Health at the University of Washington. He is a medical geographer whose specialty is infectious diseases and society, disease ecology, and health care delivery. He served as director of the university's Undergraduate Program in Public Health. He teaches classes on the geography of infectious diseases at local, national, and international scales; the environmental, cultural, and social explanations of those variations; and comparative aspects of health systems.

Linda Mearns is a scientist and deputy director of the Environmental and Societal Impacts Group of the National Center for Atmospheric Research. She holds a Ph.D. in geography/climatology from the University of California at Los Angeles. Her research focuses on climate change impacts and variability on the biosphere, particularly agroecosystems, land surface/atmosphere interactions, and analysis of climate variability and extreme climate events in both observations and climate models. She was the lead author of several chapters in the Intergovernmental Panel on Climate Change working groups I and II on subtopics dealing with regional climate and similar types of climate scenarios.

Roger Pulwarty is a research scientist at the Climate Diagnostics Center in the Cooperative Institute for Research in Environmental Sciences at the University of Colorado, Boulder. He is currently on leave, serving with National Oceanic and Atmospheric Administration-Office of Global Programs as the program manager for regional integrated assessments. He is interested in climate and its role in society/environment interactions, including the assessment and management of climate-related risks. His research has focused on factors influencing societal and environmental vulnerability to climate variations and abrupt changes and the practical utilization of climate/weather information in the western United States, Latin America, and the Caribbean. He serves on the Applied Climatology and the Societal Impacts Committees of the American Meteorological Society and is on the editorial board of the journal *Climate Research.*

Leslie Real is Asa G. Candler Professor of Biology at Emory University. He has a Ph.D. from the University of Michigan. His research interests include theoretical and evolutionary biology, population ecology and genetics, and the ecology and evolution of infectious diseases. Dr. Real has served on the National Research Council's Board on Environmental Studies and Toxicology and is currently a member of the NRC's Committee on Hormonally Active Agents in the Environment. Dr. Real also serves on the Science Advisory Board of the Environmental Protection Agency.

Chester Ropelewski is director of climate monitoring and dissemination for the International Research Institute for Climate Prediction at Columbia University. Previously, he served as a research meteorologist with the Climate Prediction Center of the National Weather Service, directing research and operational climate monitoring from 1990 to 1997. His primary research interests include studies of the El Niño/Southern Oscillation and its influence on rainfall and temperature, analysis and display of climate information, influence of land surface on atmospheric processes, and detection of global climate change. He has been a contributor to national and international reports, including the Intergovernmental Panel on Climate Change, and currently chairs the American Meteorological Society's Climate Variations Committee.

Joan Rose is a professor in the Department of Marine Science at the University of South Florida. She is currently a member of the National Research Council's Water Science and Technology Board and has served on other NRC committees dealing with water supply systems. Her area of expertise is water pollution microbiology, and her research focuses on surveys of waste waters, source waters, and drinking waters for waterborne disease-causing agents, specifically viruses and protozoa, and new technologies for the recovery, detection, and identification of pathogens in water. She was recently invited to join the National Drinking Water Advisory Council.

Robert Shope is professor of pathology at the Center for Tropical Diseases, University of Texas Medical Branch, in Galveston. He is a virologist/epidemiologist and former director of the Yale Arbovirus Research Unit. He was a member of teams that investigated outbreaks of Rift Valley fever, dengue, St. Louis encephalitis, Lyme disease, and other vectorborne diseases that are climate sensitive. He also has expertise in the diagnosis and rapid identification of human pathogenic viruses carried by arthropods and rodents. In 1992 he co-chaired the Institute of Medicine's study on emerging infections and has served on several National Research Council committees.

Joanne Simpson is chief scientist for meteorology at the National Aeronautics and Space Administration's Goddard Space Flight Center. Previously she was head of the Severe Storms Branch at Goddard, principal investigator for NASA's Tropical Rainfall Measuring Mission. She has held professorships at several universities. Her expertise spans the fields of tropical meteorology, remote sensing, atmospheric physics and engineering, Earth system science, and oceanography. She currently serves on the National Research Council's Board on Atmospheric Sciences and Climate and previously was on the NRC's Advisory Panel for the Tropical Ocean/Global Atmosphere Program. In addition, she is a member of the National Academy of Engineering, has served as president of the American Meteorological Society, and has served as a member of numerous advisory panels for American Meteorological Society, National Science Foundation, NASA, and other organizations.

Mark Wilson holds a dual appointment as associate professor in the Departments of Biology and Epidemiology at the University of Michigan. After earning his Sc.D. from the Harvard School of Public Health, he was a postdoctoral fellow at Harvard; worked at the Pasteur Institute in Dakar, Senegal; and from 1991 to 1996 was a member of the faculty at Yale University. Dr. Wilson's research addresses patterns and processes in disease ecology, particularly of human pathogens that are arthropod-borne or zoonotic. His studies of transmission dynamics, vector-host-parasite evolution, and environmental variation are directed at various viral, bacterial, and protozoal diseases and employ field studies, laboratory experiments, and modeling, including use of satellite images and geographical information systems. The goals are to reduce the risk of emerging diseases, to design ecologically sound development, and to understand the health impacts of global environmental change.

B

Speakers/Presentations at the Committee Meetings

Meeting #1: March 23-24, 1999

Donald Burke (committee member), Johns Hopkins School of Public Health
Epidemic Forecasting: Toward a Predictive Science of Emerging Infectious Diseases

Ann Carmichael (committee member), Indiana University
Historical Overview of Infectious Diseases and Society

Jonathan Patz, Johns Hopkins University
Overview of Current Climate and Health Assessment Activities

National Aeronautics and Space Administration (NASA), Earth Science
      Enterprise, Nancy Maynard
National Aeronautics and Space Administration (NASA) Life Science
      Division, David Liskowski
National Oceanic and Atmospheric Administration (NOAA), Juli Trtanj
Environmental Protection Agency (EPA), Anne Grambsch
Center for Disease Control and Prevention (CDC), Duane Gubler (*written briefing*)
National Institutes of Health (NIH/NIAID), Kate Aultman
United States Global Change Research Program (USGCRP), Robert Corell
National Science Foundation (NSF), Robert Corell
United States Geological Survey (USGS), Stephen Guptill
Overview of Relevant Federal Agency Research Activities

Meeting #2: July 20-21, 1999

John Harte (committee member) University of California, Berkeley
Scaling: From Manipulated Plots to Landscape and Regions

Roger Pielke, Jr., National Center for Atmospheric Research
The Science of Prediction and Forecasting

Paul Epstein, Harvard Medical School, Center for Health and the Global
Environment
Discussion of NRC Report Making Climate Forecasts Matter

Panel Discussion on Predictive Mathematical Modeling:

Pim Martens (committee member), Maastricht University
Models of Vectorial Capacity

Dana Focks (committee member), United States Department of Agriculture
Center for Medical, Agricultural and Veterinary Entomology
Dengue and Lyme Models

Eileen Hofmann, Old Dominion University
Modeling Studies of Oyster Disease in Chesapeake Bay

Mercedes Pascual, University of Maryland Biotechnology Institute
Nonlinear Time-Series Models

Panel Discussion on Examples of Early Warning Systems

Will Whelan, U.S. Agency for International Development; Eric Wood,
U.S. Geological Survey
Famine Early Warning System

Ken Linthicum, Walter Reed Army Institute; Compton Tucker, National
Aeronautics and Space Administration
Warning System for Rift Valley Fever

Linda Mearns (committee member), National Center for Atmospheric
Research
Agricultural Warning Systems

Joan Rose (committee member), University of South Florida
Water Quality Warning Systems

David Cleaves, U.S. Forest Service
Forest Fire Early Warning/Forecasting System

Meeting #3: October 11-13, 1999

Mark Wilson (committee member), University of Michigan
Analytic Approaches to Studying Climate Impacts on Health

Stuart Bretschneider, Syracuse University
Forecasting Methods Used in Business/Economic Disciplines

Mervyn Susser, Columbia University
Ecological Epidemiology

*Panel Discussion on the Ecology, Epidemiology, and Climate Linkages of
Specific Diseases:*

Rita Colwell, National Science Foundation
Cholera

Nancy Rosenstein, Centers for Disease Control and Prevention
Meningococcal Disease

Cynthia Lord, Florida Medical Entomology Laboratory
St. Louis Encephalitis

Gregory Gurri Glass, Johns Hopkins School of Public Health
Hantavirus

Ned Walker, Michigan State University
Lyme Disease

Meeting #4, Jan. 31- Feb. 1, 2000

William Sprigg, University of Arizona
*Overview of Planned World Meteorological Organization/World Health
Organization Conference on Climate and Health*

*Panel Discussion on the Ecology, Epidemiology, and Climate Linkages of
Specific Diseases: (continued)*

Ed Kilbourne, New York Medical College
Influenza

Steve Lindsay, University of Durham
Malaria

Paul Reiter, Center for Disease Control
Malaria and Other Mosquito-Borne Diseases

*Panel Discussion on Creating Effective Disease Early Warning Systems: The
Needs of the User Communities:*

Baruch Fischhoff, Carnegie Mellon University
Communication to the Public About Health Risks

James Leduc, Center for Disease Control
National/International Public Health Interventions

Joan Mulcare, San Bernardino County Health Department.
State/Local Public Health Interventions

Meeting #5: April 10-11, 2000

Jonathan Patz, Johns Hopkins University
The U.S. National Assessment Health Sector Report

William Lyerly, U.S. Agency for International Development
Natural Disasters and Infectious Disease Outbreaks

Steve Ostroff, Centers for Disease Control and Prevention (CDC)
Infectious Disease Surveillance Systems and Epidemiological Databases

Joel Gaydos, Walter Reed Army Institute
The DOD Global Emerging Infections Surveillance and Response System

Index

St. Louis encephalitis, 46, 49-50, 61,
   89-90
West Nile virus, 90, 93-97
yellow fever, 1, 9, 78

W

Waterborne diseases and water treatment, 7,
   38, 42, 44, 46, 79
   see also Flooding; Sanitation; Vibrios;
      Wetlands
   cholera, 16-17, 34, 38-39, 57-58, 70-71,
      79
   cryptosporidiosis, 1, 33, 46, 56-57
   dengue fever, 25, 42, 46, 47
   drought and, 38-39
   early warning systems, 90, 94, 96
   historical perspectives, 14, 16, 18, 44

malaria, 42, 46
response strategies, 90, 94
Rift Valley fever, 42
risk assessment, 38, 68
schistosomiasis, 39
temperature factors, 34
vibrios, other than cholera, 58
West Nile virus, 90, 93-97
Wetlands, 79
Wind, 12, 14, 20, 21, 34, 38, 48
World Health Organization, 9, 43, 54, 73, 134
World Meteorological Organization, 75, 134
World Weather Watch, 75

Y

Yellow fever, 1, 9, 78